Reference Sourcebook for Reducing Greenhouse Gas Emissions from Transportation Sources

February 2012

Prepared under FHWA Project DTHF61-09-F-00117

RAND Corporation
Nidhi Kalra, Liisa Ecola, Ryan Keefe, Brian Weatherford and Martin Wachs

RSG, Inc.
Peter Plumeau, Stephen Lawe, Colin Smith

Preface

About this Document

There is general scientific agreement that greenhouse gas GHG emissions are contributing to a long-term warming trend of the earth, and there is an increasing realization that transportation, as a significant contributor of GHGs, plays an important role in climate change policy and program decisions. Since 1990, transportation has been one of the fastest-growing sources of GHGs in the United States. In fact, the rise in transportation emissions represents 48% of the increase in total U.S. GHGs since 1990. In 2009, the transportation sector directly accounted for about 27% of total U.S. GHG emissions, making it the second largest source of GHG emissions, behind only electricity generation (33%). Nearly 97% of transportation GHG emissions came through direct combustion of fossil fuels.[1]

The prospect of global warming and increased climate variability has become a major policy issue during the last decade. Since transportation is a major—and growing—contributor to GHG emissions, transportation agencies will increasingly seek ways to address it by developing ways to mitigate GHG emissions. This will be especially challenging because agencies simultaneously face reduced revenue, increased congestion, and growing demands for transportation. Therefore, agencies will need guidance and information in order to meet climate change mitigation goals amid these other challenges.

This report, sponsored by the Federal Highway Administration (FHWA), helps address that need. It presents the results of a literature review of GHG mitigation strategies, summarizing what has been published about the GHG effects of different strategies, their costs, and the social feasibility of implementing them. This report does not endorse or recommend particular strategies and did not involve a direct analysis of strategies; therefore, it is best thought of as a sourcebook of information. This information can be used by transportation agencies—principally Departments of Transportation (DOTs) and Metropolitan Planning Organizations (MPOs)—to inform decision-making about strategies in their own jurisdictions. This document may also be of interest to other government agencies, researchers, transportation consultants, and students.

---

[1] http://climate.dot.gov/about/index.html, accessed 9/9/11.

Table of Contents

Preface i

1. Introduction 1

2. Summary of Findings 9

3. Research Methodology 19

4. Land Use and Transportation 25

5. Transportation Demand Management Strategies 31

6. Transportation System Management Strategies 103

7. Vehicle Improvement Strategies 171

1. Introduction

There is general scientific agreement that GHG emissions are contributing to a long-term warming trend of the earth. In the United States, rising seas already place coastal cities at risk, changing temperature and precipitation may alter the nation's food production capabilities, and increasingly extreme weather will take a toll on lives, the economy, and infrastructure (U.S. Global Change Research Program, 2009).

Yet the most severe changes in climate can still be avoided if greenhouse gas (GHG) emissions are curbed significantly. Successfully mitigating GHG emissions will require large and often difficult changes in all sectors of society, including transportation. From 1990 to 2009, GHG emissions from transportation increased 16%, and, today, transportation accounts for nearly one third of U.S. GHG emissions (U.S. Environmental Protection Agency, 2010). Thus, it is urgent that the transportation community takes steps today to reduce GHG emissions from this sector.

Some states have already implemented policies recommending or requiring that transportation agencies address climate change. California, for example, requires the California Air Resources Board to set targets for GHG emissions reductions. It further requires that Metropolitan Planning Organizations (MPOs) develop roadmaps for achieving those targets as part of their long-range transportation plans.[2] New York State's 2002 Energy Plan recommended that the state's Department of Transportation (DOT) and MPOs estimate the GHG emissions that would result from their long-range transportation plans (New York State Energy Plan, 2002). New York's 2009 Energy Plan calls for statewide GHG emission reductions to 80% below 1990 levels by 2050 (New York State Energy Plan, 2009).

As concerns over climate change grow, transportation agencies will increasingly seek ways to address it by assessing GHG emissions, setting targets for reductions, and developing ways to mitigate emissions and meet those targets (Grant et al., 2010). This will be especially challenging because federal, state, and local agencies simultaneously face reduced revenue, increased congestion, and growing demands for transportation. Therefore, agencies will need guidance and information in order to meet climate change mitigation goals amid these other challenges.

This sourcebook helps address that need. It presents the results of a literature review of GHG mitigation strategies, summarizing what has been published about the GHG effects of different strategies, their costs, and the social feasibility of implementing them. These results can be used by transportation agencies—principally DOTs and MPOs—to inform decision-making about strategies in their own jurisdictions.

This sourcebook is organized into seven chapters. Chapter 1 reviews the goals of this work, provides background on GHG emissions in transportation, and lists the strategies reviewed. Chapter 2 summarizes key findings about the strategies and offers recommendations to the FHWA

---

[2]California Sustainable Communities Planning Act.

U.S.Department of Transportation
Federal Highway Administration

about providing guidance to DOTs and MPOs. The research methodology is presented in Chapter 3 and the role of land use is presented in Chapter 4. Finally, Chapters 5, 6, and 7 present reviews of transportation demand management strategies, transportation system management strategies, and vehicle improvement strategies, respectively.

Goals of this Sourcebook

GHG mitigation strategies are policies or actions that can be used to curb greenhouse gas emissions. This sourcebook has two goals. The first is to provide DOTs and MPOs with a rich context in which to consider and evaluate GHG mitigation strategies for possible implementation in their jurisdictions. To this end, this review helps them answer the following key questions about the set of strategies as a whole:

- What is the overall state of the literature on GHG mitigation strategies?

- What conclusions can be drawn about the effectiveness of strategies? Which strategies appear more or less promising?

- What crosscutting issues exist across strategies that agencies should consider?

These questions are answered in Chapter 2. The review also addresses questions about individual strategies:

- What is the GHG mitigation strategy and how is it intended to affect emissions?

- Where it has been assessed, what has been the experience of jurisdictions that have implemented a particular strategy, in terms of:
    - The reductions of emissions?
    - The costs to agencies, the public, and other stakeholders of implementing the strategy?
    - Social concerns, such as equity, impact on mobility, and economic effects?
    - Other positive and negative consequences?
    - How the strategy interacts with other GHG mitigation strategies?

- Where these strategies have not been implemented or assessed in practice, what can be known about the strategy from theoretical research or expert judgment?

- When the literature about the strategy is considered as a whole:
    - What do we know with confidence?
    - What are the most important assumptions, uncertainties, and caveats?

- What other practical resources are available for agencies considering the strategy?

The review of individual strategies in Sections 5-7 answers these questions. As described in Chapter 3, our approach to documenting each strategy is designed to help agencies quickly obtain the answers to each of these questions.

The second goal is to highlight for FHWA, DOTs, MPOs, and other stakeholders the most pressing needs and promising opportunities for future research. As climate change gains recognition nationally and internationally as a critical concern of our time, the demand for information will grow, and the research and practice of GHG mitigation strategies will accelerate. This sourcebook is designed to help shape those efforts by identifying gaps in the current research related to each strategy that could be filled by near-term research (described in individual reviews of strategies in Chapters 5-7). Where possible, it identifies longer-term research opportunities as well, though a comprehensive assessment of research gaps for each strategy is beyond the scope of this review. This sourcebook further discusses crosscutting research needs in Chapter 2 on summary findings.

GHG Emissions in Transportation

In transportation, vehicles' consumption of fuels—most often petroleum products—is the key source of GHG emissions, and most GHG mitigation strategies target emissions from these mobile sources. The quantity of emissions from mobile sources is a product of three factors: the carbon content of the fuel, the vehicle's fuel consumption per mile of travel, and the miles the vehicle traveled.

The *carbon content* of the fuel refers to the amount of carbon that is released into the atmosphere when a quantity of that fuel is consumed. Some fuels have higher carbon content and thus produce more emissions than others. Gasoline, for example, emits 19.6 lbs of CO_2 per gallon, while diesel emits 22.4 lbs of CO_2 per gallon.[3]

The vehicle's *fuel economy* is the number of miles the vehicle can travel on a particular quantity of a particular type of fuel.[4] In 2010, the average fuel economy of all light-duty vehicles (cars, minivans, sport utility vehicles, and pickup trucks) in the U.S. was 22.5 mpg (U.S. EPA (2010), p. iii), while the average fuel economy for heavy trucks and buses was in the single digits (US Bureau of Transportation Statistics, 2009), tables 4-11 and 4-12.

---

[3] Federal Register / Vol. 75, No. 88 / Friday, May 7, 2010: Light-Duty Vehicle Greenhouse Gas Emission Standards and Corporate Average Fuel Economy Standards; Final Rule.

[4] The terms *fuel economy* and *fuel efficiency* are often used interchangeably. However, these are reciprocal terms: *Fuel economy* refers to the distance a vehicle travels per unit of fuel, while *fuel efficiency* refers to the amount of fuel needed to travel one unit of distance.

Importantly, the fuel efficiency of a vehicle is not constant but varies based on driving and maintenance habits, traffic conditions, and other factors.[5] A well-maintained vehicle is likely to have higher fuel efficiency than one of an identical model and year that is poorly maintained. Additionally, most vehicles have different fuel efficiencies at different speeds and over different terrain.[6]

The third factor is the number of miles traveled by the vehicle, or *vehicle miles traveled* (VMT).[7] The more miles the vehicle travels, the more fuel it must consume, and thus the more GHGs it emits.

Equation 1.1 is used to determine emissions based on these factors:

(Eq. 1.1) $$Emissions = CarbonContent \times \frac{1}{FuelEconomy} \times VehicleMilesTraveled$$

Therefore, for example, driving a gasoline-powered car with a fuel economy of 21 mpg for 30 miles emits 27.7 lbs of CO_2 as shown in Equation 1.2:

(Eq. 1.2) $$22.7 lbsCO2 = \frac{19.4 lbsCO2}{1 gallon} \times \frac{1 gallon}{21 miles} \times 30 miles$$

There are other sources of emissions in addition to fuel combustion in mobile vehicles. These include the production of construction materials like concrete; transportation system construction, operation, and maintenance; and vehicle manufacturing. This sourcebook considers these sources principally in terms of how they affect or interact with strategies to reduce emissions from mobile sources.

GHG Mitigation Strategies Reviewed in this Report

The strategies reviewed in this document were selected because they focused on actions that were or could be within the purview of DOTs and MPOs, as opposed to actions that only the federal government could undertake.[8,9] As shown below, these strategies are divided into three categories:

---

[5] The fuel economy ratings established by the EPA are measured under controlled conditions in laboratory settings. The EPA estimates account for many of the factors that influence fuel economy, such as the use of air conditioning, temperature extremes, and high-speed and aggressive driving. Ratings for vehicles of model year 2007 or earlier included one number for fuel economy. Ratings for 2008 model years use a range of fuel economies to reflect this variation.

[6] Vehicles typically have higher fuel efficiencies when traveling at constant speed in smooth traffic in comparison to stop-and-go traffic.

[7] This is in contrast to person-miles traveled, which reflects the number of miles traveled by the individuals in vehicle. For example, if a vehicle with one passenger drives one mile, both the vehicle miles traveled and the person miles traveled are the same: one mile. If the vehicle has two passengers and is driven one mile, the vehicle miles traveled is still one, but the person miles traveled is now two. When computing GHG emissions, the VMT is the key quantity.

[8] While many strategies could be additionally promoted by non-government entities (for examples, drivers could voluntarily choose lower-emission vehicles, or advocacy groups could mount public education

transportation demand management, transportation system management, and improvements to vehicles.[10]

Transportation Demand Management Strategies	Transportation System Management Strategies	Vehicle Improvement Strategies
• Road Pricing (including distance-based fees and cordoning) • Parking Management and Parking Pricing • Car Sharing • Pay-as-You-Drive Insurance • Ridesharing and HOV Lanes • Transit Incentives • Transit Improvements • Telework	• Traffic Signal Optimization • Ramp Metering • Incident Management • Speed Limit Reduction and Enforcement • Roundabouts • Capacity Expansion • Resurfacing Roads • Alternative Construction Materials	• Feebates • Scrappage Programs • Tax Incentives for Cleaner Vehicles • Heavy-Duty Vehicle Retrofits • Eco-Driving Education and Training and Dynamic Eco-Driving • Truck Stop Electrification and Auxiliary Power Units • Anti-Idling Regulations and Campaigns

This list covers many major surface GHG mitigation strategies available to transportation agencies, but it is not exhaustive for two reasons. First, the scope of this effort was modest and, given that new strategies are continually being introduced and evaluated, a comprehensive review is not possible. Instead, this review focuses on strategies that directly affect motorized transport. While strategies that enable and promote non-motorized transport—walking and cycling—are very important corollaries to changing motorized transport, this sourcebook is able to discuss them only briefly in Chapter 2 as they pertain to land use, and recommend that future literature reviews include those strategies as well. Second, for a few strategies considered, the body of literature is not yet large enough to allow for a review at this time. This sourcebook discusses such strategies briefly

---

campaigns about eco-driving), the authors of this sourcebook assessed only strategies that could be adopted at some level of government.

[9] Some strategies, like fuel taxes or vehicle fuel economy standards are beyond the jurisdiction of DOTs and MPOs. Nevertheless, they can have a significant impact on GHG emissions reductions. Transportation agencies may wish to work with legislatures to assess the impact of these measures and develop options for implementing them.

[10] This sourcebook does not review strategies aimed at improving fuels, such as low-carbon fuel standards, because these are beyond the purview of transportation agencies.

in Chapter 3 and recommends that, as research continues, these strategies be included in future reviews.

It is important to note that in some cases strategies do not appear on our list because they are better considered as components of other strategies. For example, intelligent transportation systems (ITS) can play an important role in improving the effectiveness of transportation system improvement strategies, and so this sourcebook discusses the role of ITS as part of those other strategies rather than as a stand-alone approach. Finally, although land use patterns are intimately tied to transportation, this sourcebook considers land use as a backdrop for transportation, rather than as a strategy per se. The sourcebook discusses this choice and the role of land use in Chapter 4.

Chapter 1 References

Grant, Michael, D'Ignazio, Janet, Ang-Olson, Jeff, et al. (2010). *Assessing Mechanisms for Integrating Transportation-Related Greenhouse Gas Reduction Objectives into Transportation Decision Making*, ICF International, Final Report for NCHRP Project 20-24(64).

New York State Energy Plan (2002). As of May 17: http://www.nysenergyplan.com/2002stateenergyplan.html.

New York State Energy Plan (2009). As of May 17: http://www.nysenergyplan.com/2009stateenergyplan.html.

US Bureau of Transportation Statistics (2009). *National Transportation Statistics 2009.* US Department of Transportation.

US Environmental Protection Agency (2010). *Inventory of U.S. Greenhouse Gas Emissions and Sinks: 1990–2008.*

US Environmental Protection Agency (2010), *Light-Duty Automotive Technology, Carbon Dioxide Emissions, and Fuel Economy Trends: 1975 Through 2010, Executive Summary.* http://www.epa.gov/otaq/cert/mpg/fetrends/420s10002.pdf.

US Environmental Protection Agency (February 2005). *Emission Facts: Average Carbon Dioxide Emissions Resulting from Gasoline and Diesel Fuel*, EPA420-F-05-001. As of May 17: http://www.epa.gov/oms/climate/420f05001.htm.

US Global Change Research Program (2009). *Global Climate Change Impacts in the United States.*

2. Summary of Findings

This section discusses general findings about the literature and GHG mitigation strategies as a whole, and addresses some of the questions raised earlier:

- What is the overall state of the literature on GHG mitigation strategies?

- What conclusions can be drawn about the effectiveness of strategies? Which strategies appear more or less promising?

- What crosscutting issues exist across strategies that agencies should consider?

Sections 5-7 present a review of individual strategies.

State of the Literature

GHG mitigation is a relatively recent but rapidly growing area of research. The literature on GHG mitigation strategies has started to develop recently relative to the literature on other major transportation concerns such as safety and air quality. This is because climate change itself has emerged recently as a critical transportation issue relative to these other areas.

Nevertheless, this is becoming an increasingly active area of research and there is much valuable information already about GHG mitigation strategies. The literature includes, for instance, many case studies that examine the effects of GHG mitigation strategies that have been piloted or implemented in different jurisdictions. There are also many case studies of the GHG effects of policies that were implemented to address *other* transportation concerns, but which simultaneously have an effect on GHG emissions. This includes, for instance, efforts to increase fuel economy for the purpose of national energy security, which has the co-benefit of reducing GHG emissions. In sum, for most strategies, there are some examples of practical implementation and effects.

The amount of and nature of knowledge varies across strategies. Some strategies have, thus far, only been assessed with transportation, economic, or other models, primarily because they have yet to be implemented fully in practice. For example, this is the case with tax incentives for alternatively-fueled vehicles: alternatively-fueled vehicles have only recently entered the market, and so the research community has, until recently, been unable to estimate the effect of policies designed to encourage their ownership. Strategies such as these are an important area of future research. Thus, the amount and nature of knowledge varies widely across strategies.

Findings Regarding GHG Effects

The effects of each strategy may vary widely depending on the implementation context. As in other areas of transportation such as air quality or congestion, the effects of a particular GHG mitigation strategy depend significantly upon the transportation and broader social and economic context in which the strategy is implemented. For instance, transit improvements can apply to a variety of modes and include increases to the frequency of service on existing routes, system-wide route

U.S. Department of Transportation
Federal Highway Administration

optimizations, the addition of new routes, and improvements to transit information and comfort. The effects of a particular type of transit improvement additionally depend on the state of the existing transit system, demographic and economic trends, and land use patterns. Even for strategies that are well defined, the use of roundabouts instead of traffic signals for example, the effects depend on traffic patterns in the region and may vary greatly. Thus, for almost all strategies, including those that have been studied extensively, there is a wide range of possible GHG reductions, costs, and other effects. The very same strategy may reduce emissions in one context, have no effects in another, and even increase emissions in a third. This variability cannot be reduced, but it can be better understood through additional research of these strategies in a still wide range of contexts. This also implies that transportation agencies should carefully evaluate each strategy in the context of their own jurisdiction, and agencies ought not adopt or discard strategies solely because they have or have not been effective in other areas. Finally, strategies that involve changes to infrastructure create GHG emissions due to construction, which can reduce or even negate the overall benefits of an action.

When strategies reduce emissions, the reductions attributable to individual strategies are typically modest relative to total emissions from all surface transportation sources. Another observation about GHG mitigation strategies is that most strategies will, at best, have a small impact on emissions relative to total emissions—reductions comprising a few percent of the total surface transportation emissions. This is because most strategies affect certain parts of the transportation system (some subset of drivers, of the traffic system, or of the fleet) and those effects are themselves modest. For example, telework strategies, which seek to encourage employers to enable employees to work from home, affect some portion of businesses for which telework is feasible, and some portion of the employees of those businesses. Further, for those employees who choose to telework, it may only affect their commute on certain days and have little effect on other days or on other types of travel.

Some strategies can achieve potentially significant reductions. While these strategies may be more difficult to implement, agencies should consider them and seek ways to address the associated concerns. Importantly, the effects of many strategies increase the more intensely they are implemented, and not all strategies are inherently limited to having small effects. When implemented at a high intensity, some strategies have the potential to achieve major reductions in emissions. For example, road pricing strategies include charging for driving in certain corridors at certain times of the day (usually peak hours), thus reducing demand. If the added cost is low, there may be little to no effect on the vehicle miles traveled. However if the cost is high there may be a significant decline in vehicle miles traveled and, consequently, GHG emissions on that corridor. For illustration, one can imagine if the cost were prohibitively high for almost everyone, virtually no driving would take place. Road pricing raises important concerns about equity and the social and economic impacts of limiting vehicular travel, and therefore it may be difficult to implement even at modest levels. With strategies such as this, the effects are typically limited to modest reductions in practice because of concerns about social, economic, and other impacts. While they may be challenging to implement, agencies should consider strategies that can achieve potentially significant reductions, and agencies should seek ways to address associated social and economic

concerns. For example, it has been suggested that inequity in road pricing may in part be addressed by using the revenue from road pricing to improve transit and other modes of transportation, which are primarily used by low-income drivers.[11]

The net GHG effect of some strategies is unknown—even when considered in a specific implementation context—because the strategies have multiple and complex effects that have rarely been evaluated. Most of the findings about a strategy focus on its immediate and intended effect on GHG emissions. For example, the research on road resurfacing as a mitigation strategy seeks to establish the extent to which smooth roads improve fuel economy and thus reduce GHG emissions. Similarly, research on incident management seeks to estimate how much congestion can be avoided when non-recurring incidents (like crashes and cargo spills) occur, and the corresponding reductions in GHG emissions. Both of these are key steps in determining the effects of these strategies.

Yet all strategies are complex and may have multiple, and sometimes unintended, effects that are difficult to assess and are not often accounted for in the literature. These unintended effects can reduce or negate the emissions reductions achieved by a strategy. Road resurfacing itself creates GHG emissions, and those emissions are usually not taken into account in research that assesses the effect of smooth roads on fuel economy. This means that it is not yet known whether road resurfacing as a mitigation strategy decreases, increases, or has no effect on GHG emissions. Incident management may reduce congestion initially, but some or all of those gains may be lost if driving increases as travelers discover that the transportation system has become more efficient and are induced to use it further. Thus, the overall effect of many strategies that have complex effects is currently unknown. This phenomenon of complex, multiple consequences is a crosscutting concern discussed in detail below in the section titled *Cross-Cutting Issues in Determining the GHG Effects of Strategies*.

As discussed below, some studies[12] have shown that bundled strategies can achieve substantial GHG reductions and agencies seeking to reduce GHG emissions should consider a multi-strategy approach, evaluating the interactions of strategies in the context of their own jurisdiction. The fact that most strategies at best produce modest GHG reductions implies that agencies seeking to reduce GHG emissions significantly will have to take a multi-strategy approach that simultaneously addresses different aspects of transportation emissions with a range of mechanisms. Furthermore,

---

[11] Equity is a complex issue that arises frequently in market-based strategies. In this case, for example, a counter argument is that most lower-income people still travel by automobile, and transit may not be available or sufficient to offset the impacts. The sourcebook discusses equity further in individual reviews of fuel taxes, road pricing, and other strategies.

[12] For example, see *Potential Changes in Emissions due to Improvements in Travel Efficiency-Final Report* (US EPA, 2011), http://www.epa.gov/otaq/stateresources/policy/420r11003.pdf, and *Moving Cooler: An Analysis of Transportation Strategies for Reducing Greenhouse Gas Emissions* (Cambridge Systematics, 2009), http://www.movingcooler.info/findings.

because the effects of each strategy can vary widely, agencies should evaluate options carefully in the context of their own jurisdictions, using the literature as a basis for such analysis.

Importantly, strategies that are implemented at the same time may interact with each other and increase, inhibit, or otherwise alter their effectiveness in reducing emissions. For example, anti-idling regulations, which administer penalties for idling, and eco-driving education, which encourages drivers to reduce idling among other things, may each reduce some emissions when implemented individually. When implemented together, however, the combination of the "stick" (regulations and penalties for idling) and the "carrot" (encouragement) could result in greater reductions than the sum of the effects each strategy in isolation.[13] On the other hand, strategies may have diminishing returns when combined. For example, strategies such as traffic signal optimization seek to improve system performance and reduce fuel consumption. They produce the greatest reductions when GHG emissions from driving are high; for example, when vehicles have poor fuel economy or the fuels they burn have high carbon content. Therefore, as strategies that seek to increase fuel economy such as fuel economy standards are implemented, the reductions from system optimization strategies decrease. The reductions from the two strategies together are typically higher than if only one strategy were implemented, but the absolute effect of the two strategies in combination is less than the sum of the effects of the individual strategies.[14] Thus, in order for agencies to use a multi-strategy approach, it is very important that they assess the interactions of strategies.

The literature cannot yet offer strong evidence about interaction effects. There is not yet sufficient evidence in the literature to determine how strategies interact, for several reasons. There are few real-world examples in which the same bundle of strategies has been implemented in similar enough ways to generalize the effects of those bundles. Moreover, to fully understand the interaction between strategies, the effects of each strategy in isolation would ideally also be known. Yet in most real-world cases, strategies cannot be implemented both in isolation and as bundles.

Sometimes, the conditions under which strategies are implemented or the nature of the strategies themselves are not conducive to rigorous analysis. For instance, eco-driving education campaigns often use television and other media to educate citizens about the benefits of better driving practices. However, accurately attributing observed changes in the general population's driving habits to the campaign (as opposed to other causes) is very difficult. Accurately attributing the combined effect of anti-idling regulations and eco-driving together may be still more difficult.

---

[13] In other words, $C > A + B$, where A is the reduction in emissions from the first strategy in isolation, B is the reduction in emissions from the second strategy in isolation, and C is the reduction from both strategies implemented together.

[14] In other words, $C < A + B$, where A is the reduction in emissions from the first strategy in isolation, B is the reduction in emissions from the second strategy in isolation, and C is the reduction from both strategies implemented together. In this case, it is typically the case that $C > A$ and $C > B$.

It is possible to model the interactions among some strategies (e.g., ramp metering and capacity expansion, which both involve transportation system improvements), and there are studies that do this. In many cases, however, the research tools and knowledge are not yet available to do so. Some studies have sought to model the effects of policy bundles at a national level,[15] but their findings may not be immediately useful for State DOTs and MPOs because of difficulties in disaggregating the findings and applying them to a state or local level. Thus, while agencies should consider any available literature on strategies' interactions, it may not offer strong evidence and they will need to conduct their own analyses.

At this time, it is more appropriate to describe what is and is not known about strategies and the key factors in assessing them, rather than to rank or quantify them. Given the variability of each strategy's effects, knowing that each strategy is complex and may have unintended consequences, and seeing that new findings about strategies are continually emerging, it is not appropriate to quantify the effects of mitigation strategies or to rank strategies at this time. This sourcebook instead describes what is and what is not known about different strategies and provides insights about the various factors that agencies should consider when evaluating each.

Stakeholders may strongly oppose strategies that, for example, raise transportation costs, reduce transportation convenience, unfairly affect some portion of the population, are too costly to the public, create inconvenience for travelers, or increase GHGs in that particular context. These concerns may be widespread, or they may be voiced by a small but important or powerful minority. Although these barriers may be difficult to overcome, the strategies should not be discarded outright: as noted earlier, some can be very effective and agencies should seek ways of implementing them.

Cross-Cutting Issues in Determining the GHG Effects of Strategies

It is important but difficult to assess total GHG effects in order to determine the true impacts of a strategy in mitigating emissions. Almost all GHG mitigation strategies produce some emissions as a side effect, in addition to reducing emissions as intended. These net effects must be analyzed to know the *true* measure of a strategy's effectiveness in mitigating emissions. For example, road resurfacing and replacing intersections with roundabouts requires new construction and maintenance of the transportation system. In scrappage programs, consumers are encouraged to replace vehicles that have low fuel economy (typically older vehicles) with new vehicles that have higher fuel economy. The process of scrapping older vehicles produces emissions and scrappage programs result in additional manufacturing of new vehicles, above and beyond business-as-usual.

Strategies can also produce emissions in other parts of the transportation system. For example, ramp metering regulates the flow of vehicles onto highways, thereby reducing congestion and delays and reducing GHG emissions. However, ramp metering increases idling when vehicles wait

---

[15] Studies that examine bundles of strategies at a national level include *Moving Cooler: An Analysis of Transportation Strategies for Reducing Greenhouse Gas Emissions* (Cambridge Systematics, 2009), http://www.movingcooler.info/findings.

for their turn to enter the highway at the ramp meters, and, by reducing congestion, may enable drivers to travel at higher speeds that reduce fuel economy. Both of these effects increase GHG emissions (in comparison to not idling or driving slower) and are unintended consequences of ramp metering.

These added construction, manufacturing, idling, and speed effects create emissions that offset some of the sought-after reductions from the strategy—fuel economy gains from smoother roads or newer vehicles, or smoother traffic flow from roundabouts and less congested highways. These unintended or secondary consequences may reduce or negate the gains from the intended effects of a strategy, meaning that a strategy may reduce, have no effect on, or even increase emissions. For example, some studies have found that the unintended emissions from ramp metering outweigh the emissions reductions, and therefore the strategy may not be effective at mitigating emissions (Cambridge Systematics, 2001).

Life-Cycle GHG emissions analyses seek to determine the net GHG emissions from all effects attributable to a product or process, including these unintended emissions. Life-cycle emissions are the true measure of a strategy's effectiveness in mitigating emissions and combating climate change. Importantly, life-cycle emissions are usually not considered in assessments of GHG mitigation strategies in the literature. In some cases, this is because the focus of a study is on establishing the intended effects (e.g., the relationship between smooth roads and increased fuel economy). In many cases, however, life-cycle emissions are extremely difficult to estimate because the second, third, and n^{th} order effects of strategies are difficult to trace and even more difficult to quantify.

By the very definition of the phrase "unintended consequences," any strategy may produce significant emissions that are not currently evident. Therefore, all strategies should be assessed carefully and second and third order effects should be traced to the extent possible. In Column E, the table highlights strategies that are *known or expected to have large unintended consequences*, where the strategy's life-cycle effect is often to have no change in emissions or even to increase emissions. These strategies have nevertheless been included in this sourcebook because transportation agencies have considered them as a way to reduce GHGs, and because they may in some circumstances be effective. When considering these strategies in particular, agencies should carefully take into account the known indirect effects.

It is important to assess the possibility that gains from some strategies—those that make transportation by roadway faster, easier, or less costly—may be reduced or lost to induced demand or rebound effects. There is an economic phenomenon that when the price of a good decreases, perhaps because the supply increases, consumption of that good increases. This phenomenon affects GHG emissions in transportation in two ways. First, policies that reduce highway congestion increase the supply of transportation. The benefits of these policies may be partially offset by additional driving that occurs in response to the improved travel conditions. This additional driving is known as "induced demand" and can be an important consideration in estimating the travel and emissions impact of traffic congestion management and other transportation system improvements.

The most often-cited example of induced demand is highway capacity expansion. When new highways are built or new lanes added to existing highways, emissions might initially decrease as congestion on other routes decreases. However, this improved efficiency decreases the cost of road travel in terms of travel time and fuel spent. This lower-cost capacity is often quickly used by new users (e.g., those who were previously not traveling or those who were using other modes) (Leeming, 1969). The net effect may be that more vehicle miles are driven than before the expansion occurred, and therefore the strategy could *increase* GHG emissions overall.

Capacity expansion is one cause of induced demand. However, any strategy that makes the transportation by roadways faster or easier—by reducing the number of vehicles on the roadways or improving transportation system performance—can potentially induce demand. Strategies that are vulnerable to induced demand are noted in Column F. While capacity expansion often results in a net *increase* in emissions, most strategies do not create new capacity. Therefore, for most strategies, induced demand may reduce or negate GHG reductions from the strategy, but it is unlikely to result in *greater* emissions. Additionally, many strategies that free up capacity are likely to have a very small effect (such as car sharing programs), and it is expected that the newly freed capacity may be too small to be observable and so there may be little or no induced demand. This is an area of further research. Induced land development, which can occur as a result of either new roadway capacity or new transit capacity, can also be a source of induced vehicle travel.

Because reductions in congestion brought about by these strategies can be partially offset by additional travel from drivers who are attracted to the less congested roads, careful analysis of the direct and indirect travel activity effects of a project is warranted. The induced travel is likely to come partly from changes in travel patterns (new trips and longer trips), and partly from shifts of travelers from other times of day, routes, and modes (such as transit). Accurate project evaluation must consider the impact of induced demand; otherwise, the benefits may be overestimated. Once properly accounted for, minimizing induced travel often depends on the quality of alternatives and complementary strategies for implementation. If the alternatives to traveling in congested conditions are inferior, a high time savings or price benefit is needed to change traveler behavior. In contrast, if alternatives are attractive, they are more likely to be successful, resulting in less induced demand and lower congestion.

A second phenomenon is known as the *rebound effect* and is associated with gains in energy efficiency or other mechanisms that reduce cost. When energy efficient technologies or systems are introduced, the cost of using them is less than their less-efficient counterparts, so they are used more. Thus, the sum of the energy reduction is lost to increased use. In transportation, the rebound effect is associated with increases in fuel economy: More-efficient vehicles are driven more than less-efficient vehicles, and some of the net energy benefits of switching to a more efficient vehicle are lost to the increase in use.

Importantly, induced demand and the rebound effect can be managed: the decrease in cost of driving that results from these strategies can be offset with an increase elsewhere in the system (e.g., through road pricing). In sum, even though strategies may be vulnerable to induced demand

and rebound effects, the extent of the effects should be evaluated carefully because reductions may still be made, particularly when coupled with market strategies.

Opportunities for Future Research

The many uncertainties in the current state of knowledge present both a need and an opportunity for future research. The sourcebook discusses opportunities for near-term research on individual strategies in the strategy reviews; here it briefly presents examples of research needs that are relevant to GHG mitigation strategies as a whole.

The variability of strategies' effects suggests that transportation agencies would benefit from tools that enable them to estimate effects of strategies in their own jurisdictions. Another component of this larger research project is aimed at developing a GHG policy analysis tool to meet this need. Additionally, research on interactions between strategies is currently limited. As GHG mitigation strategies are implemented more frequently, new opportunities will emerge for careful study of combinations of multiple strategies. Such research would provide much needed knowledge about how bundles of strategies may work. Similarly, agencies would benefit from guidance on how to assess interactions among strategies in their own jurisdictions, and how to choose bundles of strategies for long-term implementation.

There is also great value in research that assesses the life-cycle GHG effects from different strategies and that develops tools for transportation agencies to assess life-cycle effects. Such analyses are complex, so there is also a need for methodological innovations. Similarly, research is needed on the effects of induced demand for strategies other than capacity expansion, which has been widely studied. FHWA has undertaken a research project intended to further investigate the GHG reduction potential of highway operation and management strategies, which will include the effects of induced demand. The study is scheduled to be completed in 2012.

Finally, there is increasing recognition nationally and internationally that climate change is a critical concern of our time. Correspondingly, the research and practice in GHG mitigation is accelerating and new findings are continuously emerging. The authors of this sourcebook are hopeful that many outstanding uncertainties will be resolved in time, and recognize that conclusions drawn today about the effects and effectiveness of these strategies may tomorrow be reevaluated. The current uncertainties about the strategies, coupled with the dynamism of the body of research, suggest that a summary of the literature may need to be updated frequently in order to remain current. A final area of study, then, is in the methods by which information on GHG mitigation strategies can be kept up-to-date efficiently, and how this knowledge can be delivered to the broader community of practitioners, researchers, educators, and students.

Such methods could take several forms and serve many purposes. A review could evolve as the Highway Capacity Manual has, improving over several decades as the result of voluntary contributions of improvements by users of successive editions.[16] Alternatively, it could capitalize

---

[16] Highway Capacity Manual 2000, http://www.trb.org/Main/Blurbs/Highway_Capacity_Manual_2000_152169.aspx.

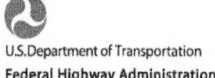

U.S. Department of Transportation
Federal Highway Administration

on today's information technologies that make such participatory efforts highly engaging and productive. One can imagine a wiki sourcebook with a large community of continuously contributing authors, an environment for discussion, and a clearinghouse for research needs and publications.[17] Such an approach could have great utility for the transportation community.

---

[17] A wiki is a website that allows users to easily create and modify hyperlinked web pages using only a common browser and with limited knowledge of syntax. Wikis are typically created to allow a community of users to contribute to a shared knowledge creation effort.

Chapter 2 References

Cambridge Systematics, Inc. et al. (2001). Twin Cities Ramp Meter Evaluation, Executive Summary. Minnesota Department of Transportation. http://www.dot.state.mn.us/rampmeter/pdf/finalreport.pdf.

Leeming, J. J. and McKay, G. M. (1969). *Road Accidents: Prevent or Punish*. Cassell.

3. Research Methodology

This literature review summarizes what is known about a wide range of individual GHG mitigation strategies and describes trends and themes across strategies. To this end, a diverse body of literature, including academic publications and journal articles; government reports at the federal, state, and local levels; and reports issued by other credible institutions were reviewed. Relevant literature was identified through searches of online databases and libraries, as well as from the authors' professional knowledge of specific topic areas. Given the modest scope of this effort, and given that it sought to report on the major findings and trends (rather than to inventory the entire literature), development of this sourcebook focused our attention on publications that aggregate and summarize findings from several individual studies. The authors also sought literature that quantified the impacts of a strategy based either on theoretical studies or on real-world implementation and assessment. The bulk of this effort was conducted at the end of 2009 and therefore the review largely focuses on the literature published prior to January 2010. That this report does not include the most recently published work is a limitation inherent to literature reviews in rapidly advancing areas of research, but one that FHWA will continue to address through supplementary updates in the future.

In addition to using the literature, the authors also drew upon their own judgment and knowledge where appropriate (e.g., in identifying likely interactions between strategies and co-benefits). In many cases, the authors also calculated the GHG or cost effects of different strategies based on data in the surveyed literature. For instance, some studies report on the reductions in fuel consumption or VMT that a strategy achieved in a particular circumstance, rather than on the GHG emissions themselves. Where feasible, the authors used data from BTS, EPA, and other sources about the average fleet size, fuel economy, carbon content of fuels, etc. to estimate effects on emissions. The authors typically used the quantities described in Table 3.1. However, sources and data varied depending upon the purposes of the calculations and the data used in each case is clearly described. The sourcebook reports small quantities of GHGs in lbs of CO_2 and larger quantities in metric tons of CO_2 (MTCO$_2$) or millions of metric tons of CO_2 (MMTCO$_2$).

To compare expenditures across projects, most dollars were converted to 2009 U.S. dollars (USD) using the Bureau of Labor Statistics (2010) inflation calculator. Where expenditures are reported in other currencies, these have been converted to 2009 USD using the OANDA conversion calculator.[18]

---

[18] This online calculator of historical exchange rates is available as of August 13, 2010 at: http://www.oanda.com/currency/historical-rates.

Table 3.1. Typical data used to estimate GHG emissions from data in the literature.

Data	Value
Carbon content of gasoline (lb CO_2/gal)	19.6*
Carbon content of diesel (lb CO_2/gal)	22.4*
Average fuel economy of all registered cars (mpg)	22.6 (2008)+
Average fuel economy of all registered light trucks (mpg)	18.1 (2008)#
Average fuel economy of all registered cars and light trucks (weighted) (mpg)	20.7 (2008)°
Average annual VMT for all registered cars (mi)	11,800 (2008)+
Average annual VMT for all registered light trucks (mi)	11,000 (2008)#
Average annual VMT for all registered cars and light trucks (weighted) (mi)	11,460 (2008)°

Sources:

*Federal Register / Vol. 75, No. 88 / Friday, May 7, 2010: Light-Duty Vehicle Greenhouse Gas Emission Standards and Corporate Average Fuel Economy Standards; Final Rule.

+Bureau of Transportation Statistics, Table 4-11 (2009).

#Bureau of Transportation Statistics, Table 4-12 (2009). Calculated from data in Bureau of Transportation Statistics, Tables 4-11 and 4-12 (2009).

In order to consistently and clearly present findings for each strategy, this sourcebook employs a standardized framework for the reviews. This framework and a brief description of each section in the framework are provided in Box 3.1. Each review essentially consists of five major sections: (1) background information and a definition of the strategy; (2) GHG, cost, and other effects; (3) concerns associated with implementing the strategy; (4) the extent to which the strategy has been widely or scarcely adopted; and (5) opportunities for new research. At the end of each review, a list of references is provided.

Finally, there are a few strategies to address GHG emissions from motorized vehicles that the authors considered initially but that were subsequently determined to not yet be ready for review according to this framework, primarily because there is not enough information available in the literature at this time. These strategies include:

- *Neighborhood electric vehicles (NEVs)*, which are plug-in electric vehicles with a maximum speed of 20-25 mph, a gross vehicle weight of less than 3,000 lbs, and a typical range of 30 miles. They are highly efficient and reduce GHGs from neighborhood trips. This sourcebook has not assessed NEVs, however, because of their limited applicability in general transportation (due to their range limitations) and because there is little research on the effect of efforts to promote NEVs.

- *Vehicle restrictions* discourage or prohibit vehicular travel in certain areas or at certain times. While the authors have reviewed certain forms of vehicle restrictions (road pricing and HOV lanes), car free streets or restrictions that allow only certain vehicles to be driven certain days were not reviewed. These have been implemented in the short term for events such as the Beijing Olympic games (Beijing Strives to Improve Air Quality as Games Draw Near, 2008), but have not yet been implemented sufficiently in long-term settings.

- *Emissions-based annual registration fees* levy vehicle registration fees according to the fuel economy of the vehicle. While there are some registration programs that vary fees by vehicle weight or offer reductions for alternatively fueled vehicles, emissions-based fees have not been implemented or sufficiently studied to allow for a review.

The authors recommend that as implementation and research on these strategies continue, they should be included in future reviews.

Box 3.1. Framework for individual strategy reviews.

Strategy Name

Each review begins with a brief overview of the strategy and summaries of GHG benefits and costs and implementation concerns.

Background

Historical context or other information necessary to understand the strategy and its effects.

Policy and Implementing Organizations

Description of the specific policy action that is necessary to implement the strategy, and the roles of various federal, state, and local agencies in implementing the strategy.

Effects

Target Group

Note of which sectors or portions of the transportation socio-technical system the strategy affects.

GHG Effects

Discussion of the sequence of effects that lead to GHG emissions reductions, what is known in the literature about each kind of effect, and findings from individual studies.

Estimated Cost per Metric Ton of CO_2 Reduction

Discussion of estimated cost per metric ton of reduction, and a demonstration and explanation of any calculations performed. Cost may reflect total public costs, agency cost, or other types of cost as appropriate.

Key Assumptions and Uncertainties

Qualitative discussion of key assumptions and uncertainties, both in findings from the literature and in findings that may be present in agencies' calculations.

Data and Tools

Annotated list of data sources, tools, and other guides that may be helpful to agencies.

Implementation Concerns

Agency Cost

Description of the kinds of costs transportation agencies may bear, and quantitative estimates of that cost where possible.

Agency Implementation Concerns

Discussion of institutional barriers or other concerns that may arise in the implementation of the strategy.

Social Concerns

Discussion of social concerns, including costs and their distribution, effects on driver behavior, etc.

Other Costs/Barriers

Discussion of costs and barriers that may not have already been articulated (e.g., costs to industry).

Interactions with Other Strategies

Annotated list of strategies that may interact with this strategy, beyond those common to all strategies in the class (TDM, vehicles, or systems).

Unique Co-benefits

Annotated list of other benefits of this strategy, beyond those common to all strategies in the class (TDM, vehicles, or systems) and not limited to transportation benefits.

Unique Negative Effects

Annotated list of negative effects of this strategy, not limited to transportation effects, beyond those common to all strategies in the class (TDM, vehicles, or systems).

Where in Use

Note of the extent to which the strategy has been implemented, and an annotated list of implementation examples.

Recommendations for Further Research

Discussion of key knowledge gaps and near-term research opportunities associated with this strategy.

References

Each strategy has its own list of references for easier use, including hyperlinks to sources where available.

Chapter 3 References

Beijing Strives To Improve Air Quality As Games Draw Near (June 23, 2008). As of May 17, 2011: http://en.beijing2008.cn/news/olympiccities/beijing/n214416867.shtml.

US Bureau of Labor Statistics (2010). Inflation calculator. As of August 13, 2010: http://data.bls.gov/cgi-bin/cpicalc.pl.

US Bureau of Transportation Statistics (2009). *National Transportation Statistics 2009.* US Department of Transportation.

US Environmental Protection Agency (2010). *Inventory of U.S. Greenhouse Gas Emissions and* US Environmental Protection Agency (February 2005). *Emission Facts: Average Carbon Dioxide Emissions Resulting from Gasoline and Diesel Fuel*, EPA420-F-05-001.

4. Land Use and Transportation

Transportation and the uses of land are intimately related. The density and mix of buildings and other features of the built environment shape people's travel needs and habits, which in turn also shape urban form. Areas of higher-density and with mixed commercial and residential buildings—known as *compact* development—are associated with greater use of modes other than personal vehicles. People walk, cycle, and use other non-motorized transport more because trip distances are typically shorter and are less likely to require travel on major roadways. Transit development and use also tends to be more feasible and desirable in compact areas, where large numbers of people can be served efficiently. When personal vehicles are used in these areas, trips tend to be shorter, and ride sharing is more feasible because there is a greater likelihood that individuals are traveling to and from similar locations (Ewing and Cervero, 2001; Cervero, 2002).

Yet, over much of the second half of the twentieth century, residential density in U.S. urban areas decreased as jobs located in city centers declined and the population living in low densities (e.g., in suburbs) increased. Areas were also often zoned for single use, either commercial or residential. While this has started to change in recent years, Ewing et al. (2008) note, "current conditions reflect the legacy of this era of sprawl."

Changes in land use patterns may be critical to reducing GHG emissions from transportation.

This raises important questions: *Can shifts in land use (e.g., toward higher densities) lead to reductions in fuel consumption and thereby reduce GHG emissions? If so, how can such shifts be brought about, and what levels of reductions can be achieved with different measures?*

Recent research suggests that changing land use patterns may indeed be a key element in reducing GHG emissions in transportation. Estimates have been made that doubling residential density across a region could reduce VMT by about 5 to 12% (Committee for the Study on the Relationships Among Development Patterns, Vehicle Miles Traveled, and Energy Consumption, 2009). A recent report examined the potential impact of land use strategies on CO_2 emissions found that, by shifting 60% of new residential growth across the United States to compact patterns, CO_2 emissions could decline by 7 to 10% from current trends by 2050, or 79 million $MTCO_2$ annually (Ewing et al., 2008).

Numerous efforts are underway to leverage the key relationship between transportation and land use to combat GHG emissions, while also creating more livable, sustainable communities. The Department of Transportation, the Department of Housing and Urban Development (HUD), and the Environmental Protection Agency (EPA) are collaborating closely on the *Partnership for Sustainable*

Communities to "help communities nationwide improve access to affordable housing, increase transportation options, and lower transportation costs while protecting the environment."[19]

At the metropolitan level, the San Diego Association of Governments (SANDAG) is addressing climate change and GHG reduction through its 2050 Regional Transportation Plan (RTP), and its Sustainable Communities Strategy (SCS) that seeks to guide the San Diego region toward a more sustainable future by integrating land use, housing, and transportation planning to create communities that are more sustainable, walkable, transit-oriented, and compact. California's Senate Bill 375 (SB 375), effective in 2009, requires each MPO in California to prepare a SCS as an integrated element of the Regional Transportation Plan. This new element shows how integrated land use and transportation planning can lead to lower GHG emissions from automobiles and light trucks. The 2050 RTP and its SCS seek to guide the San Diego region toward a more sustainable future by focusing housing and job growth in urbanized areas, protecting sensitive habitat and open space, and investing in a transportation network that provides residents and workers with transportation options to reduce greenhouse gas emissions.[20]

For several reasons, this sourcebook treats land use as a backdrop for assessing other strategies, rather than a strategy in and of itself.
For the purposes of this review—to provide DOTs and MPOs with a review of individual, actionable strategies—it is more appropriate to treat land use as a backdrop for assessing other strategies, rather than as a strategy in and of itself; this is the case for two very practical reasons. First, a change in land use (e.g., to higher density) is not itself an action but a consequence of other actions, similar to changes in zoning. Thus, when the sourcebook discusses land use as a strategy, this refers to a large and varied bundle of policies that can be used to effect changes in land use patterns. In addition to zoning, these policies include financial mechanisms encouraging growth in existing neighborhoods, near existing transit stations, and on former industrial sites (called infill and brownfield development); requirements for pedestrian and bicycle access in new developments; and siting new schools with smaller campuses in established neighborhoods.

Second, each of these actions is largely beyond the purview of transportation agencies. The authority to zone land for specific uses and densities generally resides with local governments.[21] For example, even though the evidence indicates that transit ridership may be higher when a station is surrounded by high-density development, the transit agency cannot alone ensure or enable such development: the city that controls the zoning around the station would need to allow it. Currently, transportation investments, which are generally guided by MPOs and state DOTs who

---

[19] http://www.sustainablecommunities.gov/aboutUs.html.

[20] http://www.sandag.org/index.asp?projectid=349&fuseaction=projects.detail, accessed 8/20/11.

[21] The exception is land that is unincorporated or owned by the state or federal government, in which case the county, state, or federal agency determines land uses.

control the programming of transportation funds, are often uncoordinated with land use decisions, despite that the two are intimately related. While coordination between different agencies and governments is important[22], the status quo means that land use strategies cannot be implemented by transportation agencies as, say, traffic signal optimization can be.

These two factors suggest that guidance that focuses entirely on the relationship between land use and transportation, and the ways that transportation and government agencies can work together to effect desired changes, is warranted but beyond the scope of this review. Instead, several other reports have attempted to do just this:

- The AMPO report *Noteworthy MPO Practices in Transportation-Land Use Planning Integration* reports on five MPO projects that were particularly innovative and effective in supporting the integration of transportation and land use planning, and that are transferable to other contexts (AMPO, 2004).

- The report *Driving and the Built Environment: The Effects of Compact Development on Motorized Travel, Energy Use, and CO_2 Emissions* examines the relationship between land use patterns and VMT, and how changes in land use patterns could contribute to meeting GHG emissions targets (Transportation Research Board, Committee for the Study on the Relationships Among Development Patterns, Vehicle Miles Traveled, and Energy Consumption, 2009).

- In an earlier study titled *Travel and the Built Environment: A Synthesis,* Ewing and Cervero conduct a literature review on how the built environment affects key transportation metrics (2001).[23]

Practical concerns for this review notwithstanding, there are a number of other reasons that policies to change land use must be assessed differently from strategies like traffic signal optimization, transit incentives, or eco-driving campaigns. Land use patterns are affected by other, potentially much larger forces, such as economic conditions, the social and political climate, and people's preferences in lifestyle. Moreover, the effects of these policies typically emerge slowly, sometimes over decades, and in conjunction with the effects of these other driving factors. Correspondingly, the effects on transportation may take decades to emerge and are also shaped by broader social, economic, and political trends. For all of these reasons, it is difficult to predict the effects of these actions on GHG emissions. In addition to and partly for these reasons, most of the

---

[22] Some MPOs have tried to guide land use decisions of their constituent jurisdictions, although they have no formal power to make land use decisions. For example, some MPOs have conducted exercises looking at future regional development patterns to help the public and elected officials understand the long-term potential outcomes from land-use decisions.

[23] http://pubsindex.trb.org/view.aspx?id=717403.

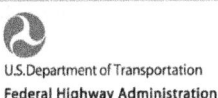

studies in this area focus on the relationship between existing land use patterns and transportation patterns, not on the effects of particular actions aimed at changing land use over a period of time.

Unlike most other strategies considered in this review, land use patterns affect much more than transportation: they can affect the balance of power between local and regional governments, the provision of utilities, housing markets and economic growth, and personal lifestyles. The effects on these segments—including on their GHG emissions and costs—should also be taken into account to understand the full effect of these actions.

These factors suggest that there are important research opportunities in assessing the effects of different actions on land use and subsequently on transportation, in developing new modeling and analyses methods that yield estimates of GHG effects, in evaluating the effects of these actions on other sectors, and in developing methods by which transportation agencies and local governments can work together on these issues.

In sum, the nexus between land use and transportation is important and may be critical to reducing GHG emissions. However, this sourcebook presents land use as a backdrop against which other strategies should be assessed, rather than as its own strategy. Where land use is known to play an important role in the outcomes of a strategy, as in car sharing and transit improvements and incentives, the sourcebook discusses the interactions in the individual strategy reviews. Conversely, where transportation strategies may have significant effects on land use patterns, as in capacity expansion, the sourcebook notes those outcomes as well.

Chapter 4 References

Association of Metropolitan Planning Organizations (2004). *Noteworthy MPO Practices in Transportation-Land Use Planning Integration.*

Cao, Xinyu, (2009). "Disentangling the Influence of Neighborhood Type and Self-Selection on Driving Behavior: An Application of Sample Selection Model," *Transportation*, Vol. 36, No. 2, pp. 207-222.

Committee for the Study on the Relationships Among Development Patterns, Vehicle Miles Traveled, and Energy Consumption, (2009). *Driving and the Built Environment: The Effects of Compact Development on Motorized Travel, Energy Use, and CO_2 Emissions*, Washington, DC: Transportation Research Board, TRB Special Report 298.

Robert Cervero (2002). "Built environments and mode choice: toward a normative framework," *Transportation Research Part D7*, 265–284.

Ewing, R., Bartholomew, K., Winkelman, S., Walters, J., and Chen, D. (2008). *Growing cooler.* Washington, DC: Urban Land Institute.

Litman, Todd. *Land Use Impacts on Transport How Land Use Factors Affect Travel Behavior.* July 9, 2010. Victoria Transport Policy Institute.

5. Transportation Demand Management Strategies

Transportation demand management (TDM) refers to a set of strategies aimed at reducing the demand for roadway travel, particularly in single occupancy vehicles. These strategies address a wide range of externalities associated with driving, including congestion, poor air quality, less livable communities, reduced public health, dependence on oil, reduced environmental health, and climate change and GHG emissions. Some TDM strategies are designed to reduce total travel demand, while others are designed to reduce peak period demand, which may disproportionately contribute to these externalities.

TDM Strategies Reviewed in This Report

This review covers the following eight TDM strategies:

Road Pricing 35

Parking Management and Parking Pricing 45

Car Sharing 53

Pay-as-You-Drive Insurance 61

Ridesharing and HOV Lanes 69

Transit Incentives 81

Transit Improvements 89

Telework 97

These strategies reduce demand through either mandatory or voluntary mechanisms. The mandatory programs reviewed discourage driving by increasing the cost of driving, as measured in money, time, or other costs. Road pricing programs charge drivers fees according to their use of the roadway, and may charge higher fees during peak periods in particular (called congestion pricing). Parking pricing charges drivers fees for parking their cars, while parking management reduces the availability of parking spaces.

Other strategies convert the fixed costs of driving in a personal vehicle into variable costs, so that the per-trip or per-mile costs are higher. As a result of the variability of trip cost, drivers tend to make fewer trips overall and VMT declines. Car sharing is a model in which participants pay to rent vehicles on a per-trip basis, and may forego owning their own vehicles. Under pay-as-you-drive (PAYD) insurance programs, drivers' premiums vary according to the miles they drive. In both cases, the *total* costs of driving can be less than they would be under the fixed-cost models. These programs generally benefit those who already drive less because they save money by paying the variable rather than the fixed costs.

TDM strategies may also make alternatives to SOV driving less expensive and more feasible. Ridesharing—meaning that more than one person travels in the vehicle—can be made more attractive by services that match drivers with passengers, provide benefits for ridesharing such as preferred parking, or operate ride sharing vehicles (e.g., corporate vanpools). High occupancy vehicle (HOV) lanes may further incentivize ridesharing by enabling ride sharers to avoid costly congestion or tolls. Transit incentives expressly reduce the cost of transit with fare passes and pre-tax payment programs, while transit improvements can increase the availability, efficiency, convenience, and comfort of transit.

Finally, strategies may reduce the need for mobility. Agencies may encourage or incentivize telework—working from home or a nearby, off-worksite location—to reduce the number or distance of commute trips.

Conclusions Regarding TDM Strategies

TDM became an important concept in transportation in the 1970s in response to the oil crises of the decade. While a number of strategies were implemented around the country at that time, most prominently ridesharing, other strategies such as car sharing have been adopted more recently. Much research has been devoted to TDM in the decades since. As noted above, TDM strategies address many externalities simultaneously, including key concerns in the 1970s of oil dependence, congestion, and air quality. These are co-benefits of all TDM strategies.[24]

The TDM strategies described above collectively reflect a "carrot-and-stick" approach: road pricing and parking pricing and management can discourage SOV driving (the "sticks") while ridesharing, transit incentives, transit improvements, and telework make alternatives to SOV more attractive (the "carrots").[25] Road and parking fees can be very effective in reducing demand, and they also generate revenue for transportation agencies. However, they are socially and economically controversial because they add to household and business expenses and the distribution of these expenses may be inequitable. These concerns are particularly important in contexts where drivers have few alternatives to SOV driving (e.g., if walking and biking are impractical or unsafe and transit availability is limited).

Conversely, making alternatives to SOV driving less expensive is typically socially acceptable because the use of those alternatives is voluntary and does not cost those who choose to drive. In part for these reasons, voluntary TDM strategies alone may only have a small effect on GHG

---

[24] These common co-benefits are cited here, and, in the individual strategy reviews, only to highlight co-benefits that are unique to each.

[25] Car sharing and PAYD individually contain both element of "sticks" and "carrots." The higher per-trip cost discourages driving, but the programs are voluntary and are likely to financially benefit those who choose to enroll. These strategies are still subject to induced demand because they do not make driving more expensive for others.

U.S. Department of Transportation
Federal Highway Administration

emissions. Importantly, these TDM strategies, and car sharing and PAYD, are also vulnerable to induced demand. When some people reduce their SOV trips, the new space on roadways makes driving less expensive (i.e., in terms of travel time), and people who previously did not make those trips or used other modes may be induced to drive. This induced demand could potentially negate some of the reductions that were initially created by these strategies.

This suggests that "carrot" and "stick" TDM strategies may be much more effective in reducing GHGs when implemented together than when either is implemented alone (i.e., they interact positively). The "stick" strategies are more effective at encouraging people to reduce SOV driving and are generally not vulnerable to induced demand because they increase the cost of driving for everyone. Simultaneously, the "carrot" strategies may provide viable alternatives to SOV driving.

Two TDM strategies aimed at encouraging alternatives to SOV driving may involve manufacturing and construction processes that produce significant GHGs. These are (1) transit improvements that involve new vehicles, new infrastructure, or increased levels of service, and (2) the creation of new capacity for HOV lanes. The GHGs produced by these processes must be considered as part of the life-cycle GHG analysis to understand the net effects of these strategies. Substantial increases in ridership or ride sharing may be necessary to produce a net reduction in emissions and these are frequently difficult to achieve, particularly in decentralized areas.[26]

As this suggests, TDM strategies and land use patterns are closely related in that compact land use is associated with lower VMT and higher incidence of non-motorized transport, ridesharing, car sharing, and transit use. In some cases, compact land use makes TDM strategies more effective, while, in other cases, TDM strategies (e.g., transit improvements) may encourage compact land use, like near new transit stops. As with land use, there is uncertainty about cause and effect: it is unclear whether people who participate in TDM programs do so because they prefer to drive less, or whether the TDM program encourages people to drive less. For example, does car sharing cause its members to drive less, or do people who already prefer to drive less choose to participate in these programs? While both phenomena are likely at work, this self-selection may limit the effectiveness of TDM programs among individuals or in areas where driving is strongly preferred.

Lastly, TDM strategies have the most significant effect on GHGs when the emissions from driving are high. For example, reducing the use of a vehicle with low fuel economy (and thus high emissions) has a greater effect than reducing the use of a vehicle with high fuel economy (and thus low emissions). This implies that TDM strategies have a reduced absolute effect when implemented with strategies that seek to reduce emissions from driving, such as fuel economy standards, fuel improvements, and transportation system improvements. All of these strategies are important in

---

[26] Transit improvements and the creation of new capacity for HOV lanes could also be thought of as transportation system management strategies as they improve the transportation system itself, in addition to decreasing demand for transportation.

U.S. Department of Transportation
Federal Highway Administration

combating climate change, but their combined effect will be less than the sum of their individual effects.

Transportation system improvement strategies and vehicle strategies may also reduce the effectiveness of some TDM strategies in another way. Improving the transportation system and improving fuel economy reduces the cost of driving, and this may induce demand and counteract the effects of transit benefits and other TDM strategies that encourage the use of alternative modes. This can be counteracted with road pricing and fuel taxes, which make up for this decrease in cost. Fuel taxes, in particular, interact *positively* with all TDM strategies because they make driving more expensive and reduce VMT.[27]

---

[27] As discussed in Chapter 7, fuel taxes and other strategies that increase the price of fuel could also be considered TDM strategies.

Road Pricing

Policy: Economic theory suggests that driving is underpriced in that current costs do not cover its significant externalities and it is thus "overconsumed." Road pricing is a market-based strategy that internalizes the costs of these externalities and facilitates reductions in total VMT or driving during peak congestion periods. Road pricing includes existing mechanisms such as toll roads, cordon pricing, and proposed approaches such as VMT charges.

Emissions Benefits and Costs: GHG effects vary depending on the form of road pricing employed and the extent of the charges. Where road pricing has been employed in practice, before-and-after studies have found that VMT was reduced by between 2 and 10%, and, where measured, GHGs declined by 2 to 6%. Some modeling-based studies have found much higher reductions, but only with very high per-mile charges that are well above the range of pricing that is normally considered in planning studies.

Implementation Concerns: While road pricing has been implemented in various forms both in other countries and in several corridors in the U.S., it remains controversial because of equity concerns, resistance to new taxes and fees, sometimes limited transit options, and privacy concerns. Simultaneously, road pricing can serve as a major revenue source, and some states are pursuing road pricing to counteract the effects of declining fuel tax revenues.

Background

According to economic theory, people overconsume goods that are underpriced. In transportation, driving is underpriced in that current costs do not cover the externalities of driving (such as pollution, GHG emissions, crashes, and congestion). As a result, driving is "overconsumed" and these externalities have become significant. Road pricing, like other market-based strategies, seeks to correct this imbalance by internalizing some of the cost of these externalities and reducing total VMT or reducing driving during periods of peak congestion. In road pricing, drivers are charged fees based on their consumption of the roadway (in contrast to fuel taxes, which charge based on the consumption of fuel). The most well-established form of road pricing in the US is toll roads, in which drivers are charged a flat fee for traveling over some section of the road, but other more sophisticated pricing schemes have also been developed.

Policy and Implementing Organizations

Road pricing can be implemented in many forms. Because the externalities of driving (like congestion) can vary based on the particular roadway used, the time of day, and other factors, a perfect road pricing system would include a real-time charge on every mile driven based on the cost of externalities produced at that particular moment. For now, this remains technologically out of reach. However, in addition to toll roads, there are three newer types of road pricing:[28]

---

[28] There are variations on all three systems, but the focus is on these three categories for this discussion.

- high-occupancy toll (HOT) lanes, in which SOV drivers can pay to enter high-occupancy vehicle (HOV) lanes;[29][30]

- cordon tolls, in which drivers pay to enter a fixed area, such as a central business district; and

- distance-based pricing, in which drivers pay by the mile driven.

In many cases, these policies are aimed at reducing congestion by enacting higher fees during peak hours; this is known as congestion pricing.

In the US, only toll lanes and HOT lanes have been implemented to date, and these have been undertaken by local or state agencies such as State DOTs. Studies of HOT lanes and cordon tolls have also been undertaken by local governments and MPOs, and pilot programs of distance-based pricing have been undertaken by several state DOTs. The federal government would play a significant role alongside state agencies in creating a national system of distance-based fees, which several panels have called for (NSTIFC, 2009; NSTPRS, 2007).

Effects
Target Group
Road pricing can target all drivers, but HOT lanes and cordon tolls typically affect only those who live or work in the targeted areas, while distance-based fees apply to most or all drivers. There may be separate rate structures for trucks versus passenger vehicles.

GHG Effects
Road pricing has been widely implemented and studied in various forms around the world. The literature on HOT lanes is mostly based on the U.S. experience, while cordon studies draw from experience in London, Stockholm, and Singapore. Several European countries use distance-based pricing for trucks, but it has yet to be implemented anywhere for all vehicles (though several pilot projects are underway). In addition to real-world studies, much research uses modeling to assess the impacts of proposed road pricing systems

Generally, the literature found evidence that people drive less, particularly with cordon systems, but this evidence comes from cities with extensive transit networks; the experience in more auto-oriented regions may be different. Although much of the literature focuses on congestion effects, some studies include GHG effects and this sourcebook focuses on those in particular. The discussion below provides results from three real-world examples of cordon and corridor pricing, as well as results from models. Because HOT lanes have largely been constructed next to free capacity, drivers

---

[29] HOV lanes are also known in some areas as carpool lanes.

[30] In some cases, the minimum occupancy for free HOT lane use may be three or more people. Cars with more than one occupant, but less than the required number of occupants, may also pay to use the HOT lane.

who do not wish to pay still have an option of driving with no charge in the same corridor. Perhaps for this reason, there is little research on the emissions impacts of HOT lanes.

In 2003, London implemented a £5 congestion charge to enter the central city during weekday business hours (from 7 PM to 6:30 PM). According to annual reports from Transport for London, the number of private cars entering London's cordon fell by almost one-third when the congestion-charging system was launched, and congestion—as measured by travel time delays—was reduced by 30%. It was additionally estimated that CO_2 emissions in the first year within the charging zone fell by 19% (TfL, 2004). When the fee was raised from £5 to £8 (from US$8.10 to $13) in 2005, CO_2 emissions decreased by an additional 5% (TfL, 2006). Interestingly, this significant decrease in vehicle trips did not correspond to a significant decrease in person trips—suggesting that mobility remained high. As one commentary put it, after the charge was introduced, there were "60,000 fewer car trips coming into the zone, [but] only 4,000 people no longer travelling to central London" (Dix, 2004). The decrease in private cars was to some extent offset by greater use of taxis, buses, and bicycles. When the charging zone was further extended westward in 2007, the reduction in CO_2 was estimated at 6.5% (TfL, 2008).

The decreases in congestion and CO_2 emissions have not consistently held over time. In 2006 and 2007, measurements of "excess delay"—the amount of additional time it takes to drive a fixed distance when roads are congested versus when traffic is free-flowing—show that congestion in London's charging zone rose slightly during those two years, back to the level in 2003 before the charge went into effect. This occurred despite the fact that the number of vehicles entering the zone remained lower than in 2003. This seems to have happened because conditions on the road network within the charging zone had deteriorated due to a number of construction projects. Based on this evidence, it seems that charging nevertheless reduced congestion from the levels it might have otherwise reached under those conditions.[31] Regarding CO_2 emissions, a later report noted, "These attributable reductions have diminished as congestion levels increased from 2006 onwards but have long since been overtaken in magnitude by the beneficial impact on year-on-year improvements to the general emissions performance of the vehicle fleet" (TfL, 2008).

In 2006, Stockholm conducted a seven-month trial of a cordon system that charged between 10 and 20 kroner ($1.46 to $2.95 in 2009 USD), depending on the time of day. Studies of the trial show that the number of vehicles crossing the cordon declined between 22% and 28% from that same time period in the previous year. When the trial ended, the volume of vehicle entries rose immediately to just below 2005 levels. Emissions fell by 41,000 $MTCO_2$ per year, or 2.7%, across greater Stockholm (City of Stockholm, 2006). A later analysis, based on data from when the congestion

---

[31] The monitoring report (TfL, 2008) drew this conclusion based on the fact that speeds at night, when the roads were not congested, had also fallen.

charging system had been re-implemented beginning in 2007, found that emissions again decreased by 2.7%, or 42,500 MTCO$_2$, per year in Stockholm County (Eliasson, 2009).[32]

Singapore's road pricing system is one of the oldest and most rigorous. It has been in place for more than 35 years and charges currently range from S$0 to S$2.50 (US$1.78 in 2009 USD), depending on the roadway and the time of day. When initially introduced, traffic on roads with charges decreased by 45% during the morning peak period, exceeding the city's 25 to 35% target. Over the course of the system's long history, congestion has not returned on those roads (Goh, 2002). Since 1998, the Land Transport Authority (LTA) has been able to maintain target speeds of 45 km/h (28 mph) on expressways and 20 km/h (12.5 mph) on arterial roads by adjusting charges as needed. This is despite the fact that the city has grown, the number of registered vehicles has increased, and no new road capacity has been added (Menon and Chin, 2004). People who no longer drive during peak hours have either moved their trips outside charging times or switched to other modes of transportation.

For comparison, in 1999, per capita emissions from road transportation in Singapore were 0.89 MTCO$_2$ while they were 5.37 MTCO$_2$ in the US. Researchers have also estimated that, at least in 1990, fuel consumption in Singapore would have been 50% higher without vehicle restraint policies (in addition to road pricing, purchase of vehicles is expensive and requires a permit). It is not clear what proportion of the 50% difference is attributable to road pricing as opposed to other vehicle restraint policies (Ang and Tan, 2001).

Several studies have also used models to assess the effects of pricing schemes in different cities.

- A study of Leeds (UK) modeled CO$_2$ emissions reductions for a variety of road pricing mechanisms. If no road-pricing mechanism was to be used (the baseline condition), CO$_2$ emissions would increase by almost 20% from 2005 to 2015. Potential CO$_2$ reductions under the other scenarios ranged from a few percent under a cordon charge of £3 (US$6.05 in 2009 USD) to about 60% with a 20-pence per kilometer (about 65 cents per mile in 2009 USD) distance-based charge. The authors concluded that a 2-pence per kilometer charge (6.5 cents per mile in 2009 USD) reduces CO$_2$ by about 12% and is probably the most desirable option: "The 10 and 20 p/km charges improve mean link speeds to above the level experienced in 1993, but the trip suppression rates are high, and are unlikely to be economically optimal, even were externalities highly valued" (Mitchell et al., 2005, p. 6238). Simply, this means that high charges lead people to forego many important trips.

- A study of Copenhagen modeled four road-pricing systems: one distance-based charge and three cordon or toll ring options, based on the empirical behavior of 500 volunteer drivers (Rich and Nielsen, 2007). In a write-up based on those results, the authors stated that this

---

[32] The City of Stockholm reported emissions in tons, and Eliasson reported emissions in "ktons," both of which were assumed to mean metric tons.

represents a reduction in CO_2 emissions of 1-3% (Rich and Nielson, 2008).[33] This reduction is lower than the London and Stockholm reductions because this study modeled a lower reduction in VMT from road pricing.[34]

- Resources for the Future modeled several road pricing schemes for the Washington, D.C. area and found reductions in "climate change costs"[35] (actual predicted values of emissions were not reported) ranging from less than 1% for a downtown cordon to 26% for a distance-based charge using social cost pricing[36] of 14.59 cents per mile (Safirova et al., 2008).

- A study of eight mid-sized cities in England modeled the effects of various levels of cordon tolls in each city. Results varied between cities as well as within cities, depending on the level of the toll. Reductions in CO_2 emissions ranged from 1.4% (a £0.75 [US$1.43 in 2009 USD] toll in Cambridge) to 14.2% (a £1.5 [US$2.85 in 2009 USD] toll in Hereford). Most were in the range of 2-4% (Santos et al., 2000).

- A study of the San Francisco Bay Area's proposed regional HOT lane network estimated it would result in 7% lower CO_2 emissions during the morning peak hours than a HOV lane network in the same corridors. The proposal seeks to convert 800 of 1,200 highway lane miles to HOT lanes (about 500 lane-miles would be converted from HOV lanes; the rest would be new HOT lanes) and to charge between 20 and 60 cents per mile in 2015 and from 50 cents to $1.00 in 2030 (MTC, 2008).

Estimated Cost per Metric Ton of CO_2 Reduction

The public cost of implementing road pricing systems (borne directly by transportation agencies) varies widely depending on the technology and the extent of the system, and can be several hundred million dollars as discussed below in the section on agency costs. Given the variation in

---

[33] This paper refers back to an earlier version of the Copenhagen study, which was published in Danish and thus not available for our review. It is not clear from this paper which system(s) result(s) in the 1-3% reduction.

[34] The authors cannot fully explain why the change in demand is lower in Copenhagen than in London or Stockholm; they suggest that it may be due to a combination of varying congestion levels, the particulars of the system, and models' tendency "to underestimate the effects of road charging" (p. 272).

[35] The authors assumed that the cost of the climate change impacts of driving is 0.44 cents per mile (in 2009 USD) (originally reported as 0.35 cents per mile in the paper), based on work by other researchers. The paper does not specify what these costs include. The paper also includes estimates for air pollution, accidents, oil dependency, noise, and congestion. All of these external costs are assigned a cents/mile figure which is then multiplied by total VMT.

[36] Social cost pricing refers to prices set such that they would capture most of the externalities of driving, such as emissions and accidents, whereas congestion pricing charges only for the congestion externality (Safirova et al., 2008).

costs and emissions reductions, a unit cost cannot be determined. Importantly, the road pricing systems that have been implemented have resulted in net revenues to agencies. (In practice, it is not likely that an agency would undertake a pricing program with costs that exceed the revenues to be generated unless the program is supported, for instance, by the federal government, through subsidies.)

Key Assumptions and Uncertainties

Many factors affect the estimated emissions impacts of road pricing. As the preceding discussion shows, measured or estimated reductions range from nearly zero to over 25%. One critical factor is the form of road pricing used. Distance-based fares generally produce greater reductions than cordon tolls or HOT lanes because the former are applied widely and at all times, while the latter are limited to certain areas or times. A second factor is the charge: the greater the charge, the greater the impact. Of course, the higher the charge, the higher the likelihood that it will be difficult to address social and economic inequities. As Safirova et al. (2008) note in their abstract, "We also find that full social cost pricing requires very high toll levels and therefore is bound to be controversial."

One study also found that the "optimal toll" varies depending on the model (sometimes by several percentage points), which in turn affects the outcome on emissions (Shepherd, 2008).

Data and Tools

FHWA has a Congestion Pricing Primer series available at http://ops.fhwa.dot.gov/tolling_pricing/resources.htm. The seven volumes in the series present issues in congestion pricing to decision-makers, including definitions, benefits, technologies, and case studies.

Implementation Concerns

Agency Cost

It is not possible to develop a precise range of cost estimates since they vary so widely depending on the location, technology, and type of implementation. This section provides several real-world cost examples (all provided in 2009 USD). London's cordon charge, implemented with a technology that reads license plates, cost approximately $378 million to implement (ECMT, 2006)[37] and has annual operating costs of $244 million (TfL, 2008).[38] Stockholm's cordon charge, using both automated plate reading and a short-range communications system, was $256 million to set up (ECMT, 2006)[39] and costs $33 million annually to operate (Eliasson, 2007).[40] In Singapore, capital

---

[37] Capital costs reported in original as €130 million (€2005) plus €144 million for additional traffic management.

[38] Operating costs reported in original as £131 in 2008.

[39] Capital costs reported in original as €190 million (€2006), including costs to operate the seven-month trial.

costs for electronic road pricing were $151 million (ECMT, 2006)[41] and operating costs are $11 million (Menon and Chin, 2004).[42] Note, however, that all of these systems raise annual revenue; operating costs as a percentage of gross revenues are 48% (London), 25% (Stockholm), and 7% (Singapore) (ECMT, 2006). This wide variation is a result of the different technologies employed.[43]

Agency Implementation Concerns

Only certain types of road pricing systems have been adopted in the United States, namely toll roads and HOT lanes. A single entity, such as a state DOT or a transportation authority generally implements these. A wider form of road pricing, such as distance-based charges, would require cooperation among a broader range of agencies. There may be challenges with technology compatibility (for example, can all transponders be used with all gantries?), working with vendors, determining which road pricing system best fits policy goals, and ensuring that payment systems function well and are enforceable.

Social Concerns

Road pricing is controversial. Several high-profile proposals have been either voted down by the public (see Greco and McQuaid [2005] about Edinburgh) or turned down by elected officials (see Confessore [2008] about New York City). The idea of paying for trips that are now free (as with a cordon toll) often raises equity concerns, but whether regressive effects occur in practice depends upon the policy and context. The ability of drivers to shift modes depends on the availability of transit services and whether land use patterns support non-motorized trips. Places that have implemented cordon tolls have fairly high levels of transit service, compared to many regions in the U.S. While there are other ways to shift driving habits—such as changing the time of travel or carpooling—in regions with road pricing, many trips have shifted to transit.

Proposals to implement distance-based fees would likely involve in-vehicle equipment, which also raises privacy concerns (Sorensen et al., 2009). Privacy can be addressed through technologies that record total fees but delete the actual locations traveled, or through the creation of anonymous accounts. However, given that other types of transponder records can be provided to law enforcement, it may be difficult to convince privacy advocates that these records can be kept confidential.

---

[40] Operating costs reported in original as SEK 220 million in 2007.

[41] Capital costs reported in original as €97 million (€1998).

[42] Operating costs reported in original as Singapore $16 million in 2004.

[43] For instance, the London system relies on license plate cameras, which requires a system to check the license plate numbers against the roster of drivers who have paid by various means, while the Singapore system operates with cheaper toll-tag technology.

Outside of conventional tolls (which have been part of the roadway network for decades), the road pricing system that has met with the most popular success in the US is HOT lanes. Public opinion polling in cities with HOT lanes finds fairly widespread support, generally 60 to 70% (Douma, 2005). This acceptance is probably due to the fact that no travel options have been taken away; that is, drivers can still travel the same corridor free of charge (unlike for example cordon systems, in which free options are not available).

Other Costs/Barriers

A key cost—that borne by drivers—cannot be generalized across program schemes and regional contexts. Setting the toll level has an enormous impact on those costs, both because they determine what drivers pay and because the levels affect driver behavior—that is, the cost may encourage some drivers to forego trips, drive at different times of day, change routes, or change modes. In addition, if drivers avoid certain trips because of the expense, this can impose other types of societal costs (for example, a driver may choose not to visit a friend, or shop in a certain area, or volunteer at a hospital). While many researchers have developed models of these costs, they must be calibrated to specific proposals, charges, and locations.

Some costs may be borne by private businesses; two examples are provided here. First, businesses may be required to assume some of the costs of the equipment that would facilitate road pricing. For example, if gas stations were required to install devices that permit pay-at-the-pump collections, that cost would be borne by the station owners (Sorensen et al., 2009). Second, businesses may experience unintended consequences of road pricing. For example, if shoppers switch from shopping within a cordon-charging area to stores on the outskirts, retail businesses inside the cordon may suffer losses in sales. However, one study examined the several road pricing schemes in Europe and found little evidence of adverse economic effects, even among businesses within cordon rings (May et al., 2010).

Interactions with Other Strategies

- Transit improvements may facilitate drivers switching modes and make market strategies like road pricing more acceptable, feasible, and equitable.

- Road pricing may also increase the success of other TDM measures such as ridesharing and telework and limit induced demand.

Unique Co-benefits

- Road pricing can raise substantial revenue for transportation investments or redistribution to drivers.

Unique Negative Effects

- Road pricing can have regressive effects, depending on how it is implemented, and depending on individuals' income, place of residence, or other characteristics

Where in Use

Cordon congestion pricing was implemented in London in 2003 and in Stockholm in 2007 (after a trial period in 2006). Road pricing has been in use in Singapore since 1975, and has gone through several changes in form. HOT lanes are in use in Orange County (State Route 91), San Diego (I-15), Minneapolis (I-394), Denver (I-25/US-36), Salt Lake City (I-15), and Houston (Katy Freeway and Northwest Freeway). Distance-based fees for trucks have been adopted in Germany, the Czech Republic, Slovakia, Austria, and Switzerland.

Recommendations for Further Research

As pricing programs are implemented around the world there is a need to document the results, assess their cost effectiveness, analyze the challenges they face, and develop a database so that future programs can be based upon best practices.

References

Ang, B. W., and Tan, K. C. (2001). Why Singapore's land transportation energy consumption is relatively low. *Natural Resources Forum*, 25(2), 135-146.

City of Stockholm. (2006). Facts and results from the Stockholm trial: final version.

Confessore, N. (2008, April 8). $8 traffic fee for Manhattan gets nowhere. New York Times.

Dix, M. (2004). Central London congestion charging. Paper presented at the ECMT Transport for London - London User Charges Conference.

Douma, F., Zmud, J., and Patterson, T. D. (2005). *Pricing comes to Minnesota: Attitudinal evaluation of I-394 HOT lane project.* Minneapolis: University of Minnesota, Hubert H. Humphrey Institute of Public Affairs.

Eliasson, J. (2009). A cost–benefit analysis of the Stockholm congestion charging system. *Transportation Research Part A: Policy and Practice*, 43(4), 468-480.

European Conference of Ministers of Transport. (2006). Road charging systems - technology choice and cost effectiveness. Paper presented at the Conference on Road Charging Systems: Technology Choice and Cost Effectiveness, Paris.

Goh, M. (2002). Congestion management and electronic road pricing in Singapore. *Journal of Transport Geography*, 10, 29-38.

Grieco, M., and McQuaid, R. (2005). Edinburgh and the politics of congestion charging: Negotiating road user charging with affected publics. *Transport Policy*, 12(5), 475-476.

May, A.D., Koh, A., Blackledge, D., and Fioretto, M. (2010). "Overcoming the barriers to implementing urban road user charging schemes," European *Transportation Research Review*, 2, 53-68.

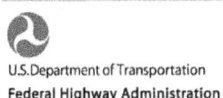

Menon, A. P. G., and Chin, K.-K. (2004). ERP in Singapore-what's been learnt from five years of operation? *Traffic Engineering and Control* 45(2), 62-65.

Metropolitan Transportation Commission (MTC) (2008). *Bay Area High-Occupancy Toll (HOT) Network Study: December 2008 Update.*

Mitchell, G., Namdeo, A., and Milne, D. (2005). The air quality impact of cordon and distance based road user charging: An empirical study of Leeds, UK. *Atmospheric environment,* 39(33), 6231-6242.

National Surface Transportation Infrastructure Financing Commission (NSTIFC). (2009). *Paying our way: A new framework for transportation finance.*

National Surface Transportation Policy and Revenue Study Commission (NSTPRS). (2007). *Transportation for tomorrow: Report of the national surface transportation policy and revenue study commission.*

Rich, J., Nielsen, O. A., and Nielsen, U. (2006). Socio-economic assessment of road pricing systems: Results from two projects in the Copenhagen region. Paper presented at the Association for European Transport.

Rich, J., and Nielsen, O. A. (2007). A socio-economic assessment of proposed road user charging schemes in Copenhagen. *Transport Policy*, 14(4), 330-345.

Rich, J., and Nielsen, O. A. (2008). External effects and road charging. In C. Jensen-Butler, B. Sloth, M. M. Larsen, B. Madsen and O. A. Nielsen (Eds.), *Road pricing, the economy and the environment.* Berlin Heidelberg: Springer.

Safirova, E. A., Houde, S., and Harrington, W. (2008). Marginal social cost pricing on a transportation network: Comparison of second-best policies (Working Paper Entry: 200803).

Santos, G., Rojey, L., and Newbery, D. (2000). The environmental benefits from road pricing (Working Paper Entry: 200404).

Shepherd, S. P. (2008). The effect of complex models of externalities on estimated optimal tolls. *Transportation*, 35(4), 559-577.

Sorensen, P., Ecola, L., Wachs, M., Donath, M., Munnich, L., and Serian, B. (2009). *Implementable strategies for shifting to direct usage-based charges for transportation funding.* Washington, DC: Transportation Research Board, NCHRP Web-Only Document 143.

Transport for London. (2004). *Impacts monitoring: Second annual report, April 2004.*

Transport for London. (2006). *Impacts monitoring: Fourth annual report, June 2006.*

Transport for London. (2008). *Impacts monitoring: Sixth annual report, July 2008.*

Parking Management and Parking Pricing

Policy: How parking is supplied, managed, and paid for varies in the US, but typically localities require developers to provide a minimum number of spaces per development type, in accordance with a formula related to the size of the development. The costs for those spaces are most often "bundled" with other development costs such that parking appears to be supplied free to drivers. This encourages driving. However, a suite of new ideas and technologies has emerged to change this paradigm, such as reducing the amount of parking and making parking more expensive. These strategies could reduce the number of driving trips and/or encourage the use of alternative modes, and many of these strategies are actually more equitable than those currently used most frequently.

Emissions Benefits and Costs: These strategies can reduce SOV trips, but they vary widely in how they are implemented, so a single range of effects cannot be generalized from the literature. In terms of costs, charging drivers for parking raises money for transportation agencies but will also make driving more expensive.

Implementation Concerns: Acceptability would likely vary by region and urban/suburban split, since many areas already have limited parking and paid parking. Both public and private parking managers might be involved, and zoning codes that govern parking requirements may need to be changed.

Background

Plentiful and free parking encourages driving. Indeed, in some cases free parking can be the main factor in the choice to drive: one study found that monthly parking charges explained up to 80% of the difference in the number of employees who drive alone to work (Dowling, Feltham, and Wyco, 1991). Moreover, virtually all vehicle trips in the U.S. have free parking on at least one end (Shoup, 2005). The goal of many parking management and parking pricing strategies is to reduce vehicle trips by making parking less available, more expensive, or both, on the assumption that people will make fewer trips, change modes, or carpool and thereby reduce GHGs.

Importantly, parking management strategies that reduce the number of spaces could create some GHGs if drivers spend significant time and fuel searching for scarce free or underpriced parking. This can be addressed in part by "smart parking" technologies, which provide real-time information about parking availability to reduce the search for parking. Simultaneously, by making parking easier, smart parking reduces some of the cost of driving that parking management and pricing strategies create.[44] The unintended consequences of both parking management and smart parking must be balanced carefully to produce a net reduction in GHGs.

---

[44] Indeed, on its own, smart parking would not be TDM strategy but a transportation system improvement strategy and could be susceptible to induced demand. Both of these effects should be considered.

U.S. Department of Transportation
Federal Highway Administration

Policy and Implementing Organizations

Parking management and parking pricing are closely related strategies. Pricing strategies charge users or owners for parking. Parking management strategies use some combination of approaches to change the amount of available parking or to require multiple users to share parking. Parking is often regulated through zoning codes that specify the minimum number of spaces that must be provided, so parking management efforts might decrease the minimum requirements, set maximum limits on parking spaces, or lower the number of parking spaces required in areas with mixed uses or near transit stations. Shared parking, on the other hand, might require that an office building make its parking spaces available in the evening to restaurant patrons.

Emerging policy ideas include "performance-managed parking" in which the availability of unoccupied spaces is maintained at 15% during peak periods through pricing, and "smart parking," in which technologies provide drivers real-time information on the availability of spaces in a particular location (whether on- or off-street).

Collectively, parking policies are typically implemented locally since cities manage their own on-street parking and set requirements for off-street parking. However, regional policies are possible.

Effects

Target Group

Parking management and pricing can be directed at the business community or individuals. Developers are the target of policies to reduce the amount of parking provided through changes in zoning or parking maximums. Other businesses may be required or encouraged to shift from free to paid parking (for example, through parking cash-out for employees), or to share parking among multiple users. Individuals can also be targeted for paid parking (for example, by charging for public garage or on-street parking). Policies to introduce paid parking for individual drivers are much more common in the U.S. than those requiring businesses to manage employee parking.

GHG Effects

There is substantial evidence from empirical studies of U.S. parking scenarios that charging for parking reduces single-occupancy vehicle (SOV) trips. Most such research focuses on commuter (work trip) parking. Studies of areas with newly-introduced paid parking (or comparisons between areas with free and paid parking) generally show that paid parking results in lower SOV mode shares, increased use of other modes, and reductions in vehicle trips. The impact on GHG depends on the number of people who stop driving alone, the emissions from the original trips (calculated based on trip length and fuel economy) and the emissions from the alternatives (whether the trip is foregone or made using another mode). Most research reports on changes in mode share and VMT and does not report on GHG effects.

Research in this area has been limited since free parking is so common and most of the "natural experiments" have been of workplace parking, so the impacts on other trips is less well understood. The elasticity of the demand for parking (that is, the change in behavior that results from a change in price) is not very high: estimates based on multiple studies have found an average of about -0.3, meaning that for every 10% increase in parking costs, the number of cars parked declines about 3% (Vaca et al., 2005). This is considered relatively inelastic, and on par with short-term elasticity for increases in fuel costs.[45]

Empirical studies of workplace parking have found that the difference in SOV mode shares is generally on the order of 7 to 15 percentage points (see various studies cited in Vaca et al. (2005)). In one oft-cited study of cash-out parking, a system in which employers allow employees who previously received free parking to receive a cash payment to stop driving alone, the average SOV mode share fell from 77% to 65% (Shoup, 1997).

It is possible to estimate impacts of paid parking on a single workplace based on the number of employees, on the amount of the parking charge (since higher costs tend to produce greater responses), and on assumptions about their willingness and ability to take fewer trips (which depends, for example, on the availability of transit and the availability of other free parking). It could also be possible to estimate such impacts at other locations, such as shopping centers, entertainment and sports venues, and hotels. However, most of the data collected on responses to parking charges are based on employees (an important demographic given that work trips are made regularly and are, on average, longer than trips for most other purposes) who tend to be more sensitive to prices than drivers with other trip purposes (Vaca et al., 2005).

Few cities have instituted widespread paid parking. Where it has been done in conjunction with existing transit service, it has been fairly successful in reducing trips. In Perth's (Australia) "parking management area," all spaces are charged an annual fee, except in residential areas and areas with fewer than five spaces. Parking charges vary depending on whether the spaces are designated short- or long-stay. These fees were first imposed in 1999. From 1991 to 2001, the percentage of center city commuters driving to work (as drivers or passengers) declined from 66% to 58%, while the percentage commuting by train increased from 5% to 18%. Over the same period, employment in the area increased from 93,000 to 97,000, indicating that the decline in mode share was not due to jobs moving outside of Perth. A major new bus route was introduced during this period, which may have played a role in this shift (Sinclair Knight Merz, 2007).

One can use some simple assumptions about fuel economy and commute distances to determine the GHG reductions from the decline in vehicle mode share that Perth experienced. If there had been no decline in mode share from 1991 to 2001, then Perth would have had 64,000 vehicle commuters in 2001. Instead, the decline to 58% means that Perth had 56,000 vehicle commuters, a difference of approximately 8,000 vehicle commuters. If one assumes vehicle occupancy of 1.15 (roughly the average vehicle occupancy over the decade [Sinclair Knight Merz, 2007]), then this results in approximately 7,000 fewer vehicles traveling into and out of the city for work. If one

---

[45] The long-term elasticity of paid parking are not currently known.

U.S. Department of Transportation
Federal Highway Administration

further assumes 250 working days a year, a 20-mile round-trip commute distance and a fuel economy of 20 mpg, this translates to an annual reduction of 1.7 million gallons of gasoline and 17,000 MTCO$_2$.[46] Although it is not possible to attribute a specific portion of the change in mode share (or subsequent GHG reductions) to parking policies alone, this figure offers an approximate upper bound on the reductions from parking policies (i.e., assuming all reductions are from parking policies, and that the alternative modes produced no additional emissions).

Smart parking is a relatively new technology, so research is still underway. One study examined smart parking at a rail transit station in Oakland, CA at which drivers could reserve spaces in advance. The study found an average VMT reduction of 9.7 miles per month per participating driver. In this pilot program, the first of its kind in the U.S., drivers could reserve spaces either online or via telephone, and changeable message signs along highways leading into downtown alerted drivers that spaces were available at the transit station. The changeable message signs were not found to be a major factor in driver behavior; only 37% of drivers surveyed who used the smart parking spaces had even seen the signs, and only one-third of those said it influenced their decision. Study results were based on a survey of drivers who used the service at least once, so conclusions were self-reported and not observed. The survey did not ask whether any participants had started driving as a result of the pilot (Rodier et al., 2008). As smart parking is still being piloted rather than fully implemented, no other studies that quantified its impacts were identified.

Estimated Cost per Metric Ton of CO$_2$ Reduction

A policy of charging for parking, which is not particularly expensive to implement compared to others, would likely result in net revenues to the implementing jurisdiction and costs to drivers. However, reductions cannot be generalized given that they are specific to policy implementation details.

Key Assumptions and Uncertainties

The greatest unknown is the response to widespread parking charging. Most U.S. studies of the response to parking charges are based on small sample sizes, such as individual worksites or parking garages, and it is unclear how these estimates would "scale up" to an entire district, city, or region.

It is also difficult to estimate the impacts of other parking management strategies, such as changing zoning codes to allow developers to provide less parking or requiring businesses to share parking. These strategies are fairly new and few assessments of their effectiveness exist. In addition, some changes in the approach to parking can take years to produce measurable changes (for example, if

---

[46] The commute distance is an educated estimate given that the 2001 average one-way commute trip length in the U.S. was approximately 12 miles (Cambridge Systematics, 2005). The fuel economy is similarly an approximation based on weighted average U.S. fuel economy of passenger cars and light trucks of 20.7 mpg in 2008 (Bureau of Transportation Statistics, 2010).

the main strategy is changes to zoning to require less parking, but development slows down, then the overall parking stock would not change very quickly).

Data and Tools
None identified.

Implementation Concerns

Agency Cost
As noted above, while the public sector would incur some costs for implementing parking management and pricing (such as collection costs, signage, enforcement, and so forth), these policies would likely produce revenue that more than covers these costs. The Perth program, for example, generated AU$9.3 million (US$8.5 million) in revenue in one year (2006-07) (quoted in Sinclair Knight Merz, 2007).

Agency Implementation Concerns
Different parking policies are implemented at different levels of government. Parking ordinances are generally enacted at the municipal level, so changing minimum parking requirements would probably have to be adopted by a city council or similar body. Policies about charging for parking may raise concerns such as how to charge for parking (for example, a previously free lot may require additional infrastructure to allow the physical means of payment) and enforce payment. Smart parking strategies may require both new policies to be adopted as well as new equipment to be procured, installed, tested, and put into service.

There can also be opposition to paid parking from groups that fear the effects of "spill-over" parking, meaning that instead of utilizing paid parking, drivers will seek out free parking. Generally the concern is that drivers will take up spaces in neighborhoods, leaving residents with limited parking options. One way to mitigate this impact is to introduce some type of permit parking, so that only neighborhood residents can park long-term in the area. This would have to be coupled with aggressive parking enforcement to be effective. These concerns may also be alleviated with more widespread parking management and pricing, so that most or all of the spaces in an area are paid or restricted in some way.

Social Concerns
Driver response may be very different depending on location, since drivers in urban areas have a wider array of travel options. They are also more accustomed to paying for parking than suburban drivers. For example, Vancouver, Canada implemented a regional parking fee that was repealed after two years in the face of continued opposition (Transport Canada, 2006). In addition, parking charges may be perceived as inequitable to low-income drivers, although this perception may not be borne out in reality.

Other Costs/Barriers

In some cases, if businesses want to pass parking charges directly to employees or customers, this may require installing equipment or technology to facilitate charging, such as adding a payment booth to a parking facility that does not currently have one. This can be overcome using a parking cash-out scheme, where employees are charged for parking through payroll deduction, unless they choose not to park a vehicle.

Interactions with other Strategies

- Parking management and pricing may be implemented alone but land use changes and transit improvements may make it easier for drivers to switch modes, thereby possibly also increasing public acceptance.

- Parking pricing may also increase the success of other TDM measures such as ridesharing and telework.

- Parking management and pricing is synergistic with car sharing: parking policies may increase the incidence of car sharing, and car sharing programs (especially with designated parking spaces) may make parking policies more acceptable.

Unique Co-benefits

- Parking pricing can raise substantial revenue for municipalities and can be used for additional transportation investments or redistributed to drivers.

Unique Negative Effects

- Parking pricing imposes costs on drivers and may have regressive effects on lower-income drivers, particularly without sufficient alternative transportation modes.

- It may also increase spill-over parking, which occurs when a shortage of parking at a particular location or for a particular purpose causes drivers to park in areas designated for other uses (e.g., residential neighborhoods or at other establishments).

- Paid parking may also affect the destination of trips (i.e., if people switch from shopping in areas with paid parking to shopping in areas with free parking. This may have adverse economic effects if the areas exist in different jurisdictions.

Where in Use

Most cities have some paid parking, both private and public, although there does not appear to be any national database that collects this information. While some Australian cities, as well as Amsterdam, use area-wide pricing (as described above in the Perth example), no American cities do so. A relatively small number of cities use parking management techniques such as performance-based pricing, reductions of parking requirements in certain areas or near transit stations, or maximum parking requirements (see, for example, Knepper et al., 2007). Many municipalities are developing smart parking programs in neighborhoods, commercial centers, airports, and other areas. San Francisco's smart parking pilot program may be the most widespread use of these technologies in the United States at this time.

Recommendations for Further Research

A closer look at examples of regional parking policies in Amsterdam and several other cities in Australia, while beyond the scope of this literature review, is likely to offer more data and information on the effects of these policies.

References

Bureau of Transportation Statistics (2009). *National Transportation Statistics, 2009.* U.S. Department of Transportation. Available online at: http://www.bts.gov/publications/national_transportation_statistics.

Cambridge Systematics (2005). *Traffic Congestion and Reliability: Trends and Advanced Strategies for Congestion Mitigation*, prepared for the Federal Highway Administration. Available online at: http://ops.fhwa.dot.gov/congestion_report.

Dowling, R., Feltham, D., and Wycko, W. (1991). Factors affecting transportation demand management program effectiveness at six San Francisco medical institutions. *Transportation Research Record*, 1321.

Hu, P. S., and Reuscher, T. R. (2004). *Summary of travel trends: 2001 National Household Travel Survey* Washington, DC: U. S. Department of Transportation, Federal Highway Administration.

Knepper, V., Wilbur Smith Associates, Michael R. Kodama Planning Consultants, Willson, R., KT Analytics Inc., Rick Williams Consulting, and CHS Consulting Group. (2007). Developing parking policies to support smart growth in local jurisdictions: Best practices. Oakland, CA: Metropolitan Transportation Commission.

Rodier, C. J., S. A. Shaheen, and C, Kemmerer. (2008, June). Smart Parking Management Field Test: A Bay Area Rapid Transit (BART) District Parking Demonstration; Final Report: California Partners for Advanced Transit and Highways Program, Institute of Transportation Studies, University of California at Berkeley, UCB-ITS-PRR-2008-5.

Shoup, D. (1997). *Evaluating the effects of parking cash out: Eight case studies, final report.* Sacramento, CA: California Air Resources Board Research Division.

Shoup, D. (2005). *The high cost of free parking.* Chicago: American Planning Association.

Sinclair Knight Merz. (2007). Review of Perth parking policy.

Transport Canada. (2006). *Urban transportation showcase program: TransLink parking tax case study.* Vancouver: Transport Canada.

Vaca, E., and Kuzmyak, J. R. (2005). Chapter 13 - parking pricing and fees. In *TCRP report 95: Traveler response to transportation system changes*. Washington, D.C.: Transportation Research Board.

Car Sharing

Policy: Most miles driven in the United States are in privately owned vehicles. Because vehicle ownership entails many "sunk costs" (e.g., the purchase price, registration fees, insurance, maintenance, etc.), out-of-pocket costs tend to be low relative to other modes on a per-trip basis, making driving attractive. Car sharing seeks to convert these fixed costs to variable ones by promoting a model in which participants rent vehicles on an as-needed basis, and may forego owning their own vehicles. As a result of the variability of trip cost, drivers tend to make fewer trips overall and VMT declines. Additionally, the vehicles available in car sharing programs are often more fuel-efficient than the average privately owned vehicle, which also reduces GHGs.

Emissions Benefits and Costs: GHG emissions have declined among car sharing members as they both drive fewer miles and more efficient vehicles. Studies in the U.S. and Canada have found that emissions declined on average by between 0.8 and 1.2 $MTCO_2$ annually per member, even after accounting for those members who drive more often because they did not previously own vehicles. No research has linked the cost of these programs to the public sector to adoption rates since most shared cars are managed by for-profit companies. Nevertheless, the public sector can play a role by providing subsidies, publicity, or parking spaces.

Implementation Concerns: While barriers to implementation are low, they may include resistance to converting public parking to parking reserved for car sharing. Importantly, car sharing has thus far been effective primarily in more compact neighborhoods or in areas with already limited parking (such as college campuses).

Background

Car ownership entails many "sunk costs" that are fixed at the same rate regardless of the amount the vehicle is driven. In a car sharing organization, members rent vehicles by the hour or day. This differs from conventional rental cars in several ways: it is marketed to residents and businesses in a city, rather than visitors; it provides hourly rates, while most rental car firms charge by the day or week; it positions vehicles throughout an area so that members can walk to them in their neighborhoods; and it emphasizes quick booking when a vehicle is needed. For some, using a car sharing service may be less expensive overall than privately owning a vehicle. For others, it may offer mobility that they would otherwise not have.

Car sharing can reduce GHGs by reducing the number of trips. Research has shown that drivers make decisions regarding modes for a particular trip based on out-of-pocket costs that vary by trip (gas, tolls, and parking), meaning that many vehicle trips in personally owned vehicles appear quite inexpensive compared with alternatives such as transit (Steininger, Vogl, and Zettl, 1996). In car sharing, these costs are variable and incurred largely per-trip, so drivers are more likely to consider the total costs and make fewer trips overall. Importantly, these programs simultaneously create a way for people who do not own their own car to drive where otherwise they may have walked, used transit, or not taken the trip. Car sharing can also reduce emissions if the vehicles in the

service have higher fuel economy than privately owned vehicles, or if members have the flexibility to choose the size of vehicle that meets their needs for each particular trip—meaning that large and less fuel efficient vehicles may be chosen only when needed.

Car sharing began in Europe and has spread to the U.S. in the past decade. Car sharing services are generally operated by commercial or non-profit entities. Members of a car sharing organization generally pay fixed fees to join and an annual membership fee, plus the hourly or daily rental fees. While car sharing organizations can have multiple locations, they tend to be most effective in high-density areas where many other trips can be served by transit or non-motorized transportation. College towns and urban university campuses have also been good markets because campuses are typically compact, students often use cars infrequently, and campus parking may be limited. Business programs in which employers join and provide car sharing as a benefit for their employees have also been growing.

Policy and Implementing Organizations

Car sharing tends to be championed by regional or local agencies, such as MPOs, local governments, and/or transit agencies. Since car sharing is largely operated by private entities, the role of the public sector may include subsidies for program start-up costs, provision of parking spaces for the vehicles, tax incentives, encouraging or requiring private developers to include car sharing spaces in multi-family housing, and publicity.

Effects

Target Group

Car sharing has been marketed to both individuals for personal travel and to businesses as a lower-cost alternative to maintaining a vehicle fleet and for employees who need access to vehicles during the work day. Various studies of total ownership costs report "break-even" points (at which the cost of car sharing equals the cost of car ownership) variously at 5,000 (Millard-Ball et al, 2005), 8,000 (Higginbotham, 2001), and 4,000 to 10,000 miles (Litman, 1999, Figure 2). Drivers who drive fewer miles than the break-even point would save money with car sharing and are potential car sharing candidates, while those who drive more are better off economically owning a vehicle and would not be good candidates.

GHG Effects

Existing car sharing programs in the U.S., Canada, and Europe have been studied to assess their effectiveness at reducing VMT and emissions among users. Among all car sharing members, both emissions and VMT decline; this reflects both the previous car owners whose emissions and VMT fell substantially, and those who did not previously own cars and now drive more. It is also possible that some emissions reductions are due to changes in the fleet mix; that is, on average shared vehicles may have lower emissions profiles than privately owned vehicles. Millard-Ball et al. (2005) observed that car sharing fleets tend to have more alternatively fueled vehicles, newer vehicles, and smaller vehicles than the overall fleet. While fleets do contain larger vehicles for special purposes (a pick-up truck to haul furniture, for example), members can choose the vehicle most appropriate for their trip. The literature does not tell us how agencies' actions (e.g., provision of subsidies and parking spaces) affect the availability and use of car sharing programs, given that it depends more on density, transit, and land use.

Only a few studies have directly estimated the effects of car sharing on GHG emissions. In a recent study based on survey responses from over 6,200 car sharing members in North America, Martin and Shaheen (2010) found that on average, a household reduces its GHG by 0.84 $MTCO_2$ per year after joining car sharing. This includes two kinds of reductions. First, it includes the "observed impact"—the observed difference between a household's actual VMT before and after joining car sharing. Two changes may be observed. Car sharing offers vehicles to people who previously did not have access, thus increasing their VMT and GHG emissions. This increase was observed in most households participating in car sharing, but the observed increase was small. The minority of households substitute a car-sharing vehicle for a personal vehicle, and significantly reduce their VMT and GHG emissions. The net effect is an overall reduction in VMT and GHG.

Second, this study assessed the avoided emissions, which are not observable. Some households joined a car sharing program instead of purchasing a new vehicle, which they would likely have driven much more than the shared vehicle. Thus, the reduction also includes the VMT avoided by households that chose not to purchase a vehicle. Together, the observed impact and the avoided emissions constitute the "full impact" of car-sharing.

The authors caution that one cannot assume every household will decrease its GHG upon joining a car sharing organization, but that the overall effect is a statistically significant net reduction. When the authors account for the inactive share of car sharing members (between 15 and 40% of all members seldom use shared vehicles), they estimate that the annual aggregate impact of car sharing reduces between 160,000 and 225,000 $MTCO_2$ per year. This seems to include the reductions that result from car sharing fleets being more fuel-efficient; the authors' data show that the vehicles the members shed after joining car sharing organizations had fuel economies of 10 mpg less on average than the shared vehicles (32.8 mpg vs. 23.3 mpg). Emissions reductions attributable to land use and vehicle production were not assessed in this study, and college and business users of car sharing were excluded.

Studies of GHG reductions from car sharing have also been done in other countries. A study found that car sharing in Quebec reduced emissions by 1.2 $MTCO_2$ per member per year. This was based on data that on average, members drove 2,900 fewer kilometers (1,800 miles) per year, and used lower-emissions vehicles (Communauto, Conseil Regional de l'environnement de Montreal, and Equiterre, 2007).[47] Ryden and Morin (2005) claimed that in Europe, car sharing reduces members' CO_2 emissions by 40 to 50%. Specifically, two estimates from European programs found decreases in GHG emissions per member of 54% (Bremen, Germany) and 39% (Belgium). Note that VMT

---

[47] The press release is cited here as the original report is not available in English.

U.S. Department of Transportation
Federal Highway Administration

reductions in these programs were higher than results from American programs. These estimates account for changes in the vehicle fleet mix, as well as increased emissions from transit use (Rydén and Morin [2005] quoted in Millard-Ball et al. [2005]).

Additional research focuses on VMT effects among car sharing members (without considering the differences in fuel economy between shared and personally-owned vehicles). Although VMT generally rose among those who had not previously owned a vehicle, those increases were generally more than cancelled out by the reductions in VMT among those who previously owned vehicles, leading to net reductions in VMT. One review of four U.S. car sharing programs found that average per-member VMT decreases ranged from 7 to 43%. Of these, two Portland, Oregon studies found average decreases of 18% in those who had previously owned vehicles, and 7.5% total decreases; neither of these findings were statistically significant. One of these Portland studies compared those who had previously owned vehicles to those who had not; VMT among owners dropped from 103 to 84 VMT per week, and VMT among non-owners rose from 0.2 to 25 VMT per week. A San Francisco study found second-year average VMT reductions of 2.8 to 1.5 VMT per weekday, and an Arlington (Virginia) study found average VMT decreases of 43%. The San Francisco and Arlington studies included but did not separately report the effects from members who previously owned vehicles from those who did not. For North American car sharing organizations, the proportion of all members who had previously owned vehicles was on average 40% (Millard-Ball et al., 2005).

In a study of nine European programs, average VMT fell between 26 and 72%. The European programs were more likely than the American ones to report changes in VMT both for members who previously owned vehicles and those who did not. For example, an Austrian study found that members who previously owned cars decreased their VMT from 10,100 to 3,850, while those previously without vehicles increased from 830 to 1,800 (Steininger, Vogl and Zettl, 1996, quoted in Millard-Ball et al, 2005). A British study found that members who previously owned a vehicle reduced their VMT by 1,100 while those who did not own a vehicle increased by 475 (Ledbury, 2004, quoted in Millard-Ball et al, 2005). On average, European car share members were evenly split between those who had previously owned a vehicle and those who did not (Millard-Ball et al, 2005).

Estimated Cost per Metric Ton of CO_2 Reduction

This is not currently known. According to Millard-Ball et al. (2005), who surveyed dozens of public entities that partnered with car sharing organizations, about 60% of car sharing organizations have received some public money for start-up costs. However, there are no existing studies that link the public expenditures to promote car sharing to the GHG reductions. It is also impossible to estimate it since there is at best an indirect link between public expenditures and GHG reduction.

Key Assumptions and Uncertainties

Estimates of car sharing effects and effectiveness are based on many assumptions, including:

- the potential membership in car sharing programs;
- the fleet mix of shared cars in comparison to that of the overall passenger vehicle fleet;
- the effect on VMT from car sharing; and
- the amount of public expenditure per current and new members.

One challenge in studying GHG reduction is that, because car sharing is entirely voluntary, it is difficult to establish a control group. Those who choose to enroll in car sharing programs may be more concerned about the environmental impact of their actions than the general public. Cervero and Tsai (2003) note that most "early adopters" in San Francisco's program were "environmentalists and avid cyclists who owned no car" (p. 44). Other studies looking at the motivations for joining car sharing found that as car sharing programs matured, the environmental rationale declined and members were more motivated by financial considerations (Harms and Truffer, [1998], quoted in Millard-Ball et al., [2005]).

In addition, because car sharing is quite new in the U.S., it is difficult to estimate the growth rates and potential of car sharing. One study estimated that 12.5% of the over-21 population in major U.S. metropolitan areas are potential candidates for car sharing (Shaheen et al., 2006). In absolute terms, this suggests a potential car sharing membership of 21 million people.[48] However, as most published research has examined car sharing in major metropolitan areas, one cannot say much about the potential impacts in smaller cities.

Finally, there does not appear to be any data source that provides the fleet mix for shared vehicles. This means that it is difficult to accurately estimate the potential reductions in emissions based on the presumed lower emissions profiles of shared vehicles.

Data and Tools
None identified.

Implementation Concerns
Agency Cost
Promoting car sharing does not require major infrastructure investments or adoption of new technologies by the public sector, and agencies' costs relative to other strategies would be fairly low. While some public agencies may provide subsidies, such support would generally come at the beginning of the venture, since car sharing organizations can operate successfully based on

---

[48] This was calculated under the assumptions that (1) major metropolitan areas are those with populations over 200,000, (2) that 233 million people therefore live in major metropolitan areas, and (3) the statistic that 73% of the total U.S. population is over the age of 20 holds true for the population living in these same metropolitan areas (U.S. Census, Population Estimates, http://www.census.gov/popest/estbygeo.html).

revenues from members. Revenues to cover program costs can also come from payments for public parking made available to shared cars.

Per Millard-Ball et al. (2005), most of these start-up grants have been under $100,000.[49] Cities have also donated vehicle parking spaces, but in the case of on-street spaces it is difficult to estimate a cost. Brookline, Massachusetts values its donated spaces at $750 per year per space, and Philadelphia, Pennsylvania charges the operator a one-time fee of $250 per space to cover the staff and signage costs. Simultaneously, some cities have substituted car sharing vehicles for city fleet cars and saved money.

Agency Implementation Concerns

Because car sharing is fairly new to the U.S., some agencies may not be familiar with it and/or may be skeptical about its viability. There may not be a natural "home" for promoting car sharing within multiple agencies. Zoning regulations may make it difficult to site car sharing vehicles (Millard-Ball et al., 2005).

Social Concerns

Because car sharing is generally voluntary, offers more options to travelers, and can be sustained by private companies from revenues, social acceptability of car sharing is generally high. There may be some public objections to using previously public parking for car sharing vehicles, or requiring car sharing parking spaces in new residential development, but this has not been significant.

Other Costs/Barriers

As car sharing entails costs to members, it may be difficult for low-income groups who could otherwise benefit from occasional access to vehicles to participate.

Interactions with other Strategies

- Car sharing programs (or efforts to encourage them) are not likely to be successful unless land use patterns and transit provisions support travel modes other than driving. Therefore, car sharing would benefit from strategies to improve transit and provide more compact land uses.

- Car sharing could reduce transit ridership.

Unique Co-benefits

- Car sharing may reduce total transportation costs for members.

---

[49] The years in which start-up subsidies were provided were not reported, so these amounts have not been converted to 2009 USD.

Unique Negative Effects
- Communities may face some loss of private parking spaces.

Where in Use
Current car sharing membership in the U.S. (as of January 2010) is about 390,000, with 7,500 vehicles (IMR, 2010). Car sharing currently operates in dozens of metropolitan areas and college towns in the U.S. A list is available at http://www.carsharing.net/where.html.

Car sharing is not limited to major metropolitan areas—Zipcar, the largest car sharing organization in the U.S., operates in several cities with populations less than 200,000, such as Winona, Minnesota, and Waterville, Maine (Zipcar, 2009).

Recommendations for Further Research
It would be useful to assess additional studies of individual car sharing programs, as well as analysis of established European programs to determine the long-term impacts of car sharing. Additionally, data on differences among members who previously owned cars versus those who did not, according to the type of city, country, economic situation, and other factors would help in assessing the promise of car sharing programs in other areas. A further area of study includes understanding how diverse fleet mixes offered by car sharing companies may affect GHG emissions.

References
Bureau of Transportation Statistics (2009). *National Transportation Statistics, 2009.* U.S. Department of Transportation. Available online at: http://www.bts.gov/publications/national_transportation_statistics/.

Cervero, R., and Tsai, Y.-H. (2003). San Francisco city carshare: Travel demand trends and second-year impacts: Institute of Urban and Regional Development, University of California at Berkeley, Working Paper 2003-05.

Communauto, Conseil Regional de l'environnement de Montreal, and Equiterre. (2007). CO_2 Emissions Reduced by 168 000 Tons Per Year Thanks to Car-Sharing, press release, Quebec City, February 19.

Harms, S. and Truffer, B. (1998). The Emergence of a Nationwide Carsharing Co-operative in Switzerland. Prepared for EAWAG – Eidg. Anstalt fur Wasserversorgung. Abwasserreinigung und Gewasserschutz. Switzerland.

Innovative Mobility Research. (2010). Carsharing. from http://www.innovativemobility.org/carsharing/index.shtml.

Ledbury, M. (2004). UK Car Clubs: An Effective Way of Cutting Vehicle Usage and Emissions? M.Sc. thesis, Environmental Change Institute, University of Oxford.

Litman, T. (1999). Evaluating carsharing benefits: Victoria Transport Policy Institute.

Martin, E.W. and S.A. Shaheen. (2010) Greenhouse Gas Emission Impacts Of Carsharing in North America: Mineta Transportation Institute.

Millard-Ball, A., Murray, G., Schure, J. t., Fox, C., and Burkhardt, J. (2005). Car-sharing: Where and how it succeeds. Washington, DC: Transportation Research Board, Transit Cooperative Research Program Report 108, Chapter 5.

Rydén, C., and Morin, E. (2005). Moses environmental assessment report.

Shaheen, S. A., Cohen, A. P., and Roberts, J. D. (2006). Carsharing in North America: Market growth, current developments, and future potential: Institute of Transportation Studies, University of California, Davis, UCD-ITS-RR-05-11.

Steininger, K., Vogl, C., and Zettl, R. (1996). *Car-sharing organizations: The size of the market segment and revealed change in mobility behavior.* Transport Policy, 3(4), 177-185.

Pay-as-You-Drive Insurance

Policy: Many auto insurance policies have fixed premiums that are based on driver demographics, driving history, vehicle type, and other factors. Pay-as-you-drive insurance (PAYD, also known as pay-at-the-pump or cents-per-mile insurance) allows drivers to purchase insurance that varies the premium based on the amount driven. This converts some of the presently-fixed costs of driving to variable costs, and drivers can save money by reducing the number of miles driven.

Emissions Benefits and Costs: Pay-as-you-drive insurance is not widespread, so there are no overall figures about how much GHG is reduced by decreases in VMT. Both modeled and empirical studies have found reductions in VMT of approximately 5 to 10% per vehicle/policy, with national modeled results showing greater reductions than small pilot projects.

Implementation Concerns: While costs to public agencies are minimal, in many states pay-as-you drive insurance is not allowed for various reasons. Some states, for example, require insurance costs to be stated at the time insurance is purchased. With PAYD, costs vary based on actual miles driven. It is unclear how quickly PAYD would be offered or used even if more widely available, or how many companies would be interested in offering it. In some cases, the technologies used to determine the number of miles driven may raise privacy or enforcement concerns.

Background

Car ownership entails many "sunk costs" such as the purchase price of the vehicle, registration fees, and insurance, which are not affected by the amount that the vehicle is driven. Some research has shown that drivers make decisions about which mode to choose for a particular trip based on immediate marginal out-of-pocket costs for the trip (e.g., gas, parking, and bus fare). This means that many personal vehicle trips appear to be inexpensive in an absolute sense, and when compared to alternatives such as transit (Steininger, Vogl, and Zettl, 1996).

Pay-as-you-drive insurance (PAYD) allows drivers to purchase insurance that varies the premium based on estimated or actual driving distances within a certain period.[50] The principle behind PAYD is that if costs vary based on vehicle use, drivers will consider the total costs and make fewer personal vehicle trips by making fewer trips or by switching to other modes. In comparison to fixed insurance rates, these options reduce GHGs and can save consumers money (Litman, 2009).

PAYD insurance can be implemented in multiple ways:

- Mileage-based discounts: the insurance company lowers the premium if drivers drive less than a certain number of miles per year or within mileage ranges;

---

[50] The rate may vary according to traditional automobile insurance factors such as driving record and age.

U.S. Department of Transportation
Federal Highway Administration

- Policies that are in effect only with a certain number of miles driven: that is, a policyholder is covered for the next 3,000 miles driven (as opposed to a time-based insurance policy that covers six months); and

- A per-mile premium: every mile driven is charged at a specific rate, which can vary based on the driving conditions (for example, a mile driven late at night on a Saturday is less expensive than a mile driven on Tuesday during commute hours).

Technologies that can be used to account for mileage include odometer readings, global positioning systems (GPS), and units that receive data from the on-board diagnostic equipment. Odometer readings can be self-reported, but verified by mechanics or the company. GPS and the on-board units can transmit data to the insurance company.

Policy and Implementing Organizations

Because insurance is regulated at the state level, whether PAYD is legal depends on state policies. Currently 34 states allow some form of PAYD insurance. However, regulations vary, so states that allow some type of mileage-based discount may not allow per-mile premiums. The GHG mitigation strategy is to legalize various forms of PAYD insurance, thereby allowing private insurance companies to develop a broad range of PAYD insurance products. Further action may be to encourage the uptake of these policies through education and publicity campaigns.

Effects

Target Group

PAYD adoption is targeted at two groups: (1) private insurance companies, which would develop and offer PAYD to policyholders; and (2) individual drivers, who would choose them over conventional insurance products.

GHG Effects

Studies involving small-scale PAYD pilot programs have been conducted in the U.S., and the literature provides estimates of the effects of PAYD if it were implemented across the U.S. There is consensus that, when PAYD is used, it decreases overall VMT among policyholders by between about 5 and 10%, although empirical studies show reductions on the smaller end of this range, in comparison to studies based on modeling. Note that studies using models assume all drivers have access to PAYD insurance; they do not make assumptions about the rate at which companies begin to offer PAYD policies or drivers switch to them.

A 2004 pilot program in Minnesota tested the response of 130 drivers who volunteered to participate (their actual insurance levels were not affected, but they were able to benefit financially if their mileage declined). Overall VMT declined by 4.4%, with greater decreases seen during weekday peak hours (6.6%) and on weekends (8.1%). Households that were willing or able reduce their VMT did so at fairly low payment thresholds (5 cents per mile), but other households did not change their mileage even at much higher levels (25 cents per mile) (Cambridge Systematics et al.,

2006). In a study of about 3,000 households in Texas who participated in a Progressive Insurance study, the average reduction in VMT was 5%. Peak hour miles were reduced by only 3.2%, which was the opposite of the Minnesota findings (Progressive Insurance and NCTCOG, 2007).

If one assumes that PAYD insurance policyholders reduce their annual VMT by 5%, one can compute the annual CO_2 reductions for cars and light trucks:

> For passenger cars: Using 2008 values, a 5% reduction from the average 11,800 VMT annually (National Transportation Statistics, 2009) to 11,200 VMT annually, and assuming 22.6 MPG (the average fuel economy for passenger vehicles in 2008 (Bureau of Transportation Statistics, 2009)), results in approximately 26 gallons of gas saved and 510 lbs of CO_2 reduced annually.

> For light trucks: Using 2008 values, a 5% reduction from the average 11,000 VMT annually (National Transportation Statistics, 2009) to 10,450 VMT annually, and assuming 18.1 MPG (the average fuel economy for light trucks in 2006 (Bureau of Transportation Statistics, 2009)), results in approximately 30 gallons of gas saved and 595 lbs of CO_2 reduced annually.

Three nationwide studies using models are widely cited in the discussion of the impacts of PAYD. All of them assume that all drivers would be using PAYD insurance. One found an overall reduction in VMT from 9.2 to10%, depending on the type of PAYD model used and the state, since states differ in their average insurance costs per mile (Edlin, 2003). A second study claims that "fully implementing" PAYD (which is not precisely defined in the paper, but presumably means that all U.S. drivers are PAYD subscribers) would save 11.4 billion gallons of gasoline per year (a savings of 9.1%) (Parry, 2005). This translates to approximately 100 million $MTCO_2$. The most recent study found that PAYD would annually reduce VMT by 8%, oil consumption by 4%, and CO_2 emissions by 2% from 2006 levels (Bordoff and Noel, 2007). This estimate is lower than the others because of fuel price increases, meaning that consumer response would be smaller since the proportion of driving costs attributable to insurance has declined.

Estimated Cost per Metric Ton of CO_2 Reduction

There are no existing studies that specifically address this question, nor even any that estimate the cost of introducing PAYD insurance. Bordoff and Noel (2008) declined to estimate costs, stating that the main cost would be to install equipment and prices for nationwide implementation are very uncertain (p. 36). They also claim that any cost would be outweighed by the societal benefits of reduced congestion, crashes, and savings to drivers, although these assertions were not included in the model.

Key Assumptions and Uncertainties

One main uncertainty in estimating the effects of PAYD programs is how quickly PAYD insurance would spread in the marketplace, once legal. Only a handful of companies currently offer policies based on miles driven, and other insurance companies do not seem to be coming to market with similar products. It is also unknown how many drivers would change to PAYD insurance. While there is a clear incentive for those who drive fewer miles than average to switch policy types, those who drive the average or above would pay the same or more (depending how the policy is structured) with PAYD, so it is not clear whether they would switch voluntarily.

Another key uncertainty is the degree to which drivers might reduce their VMT with PAYD insurance. Studies that use models to predict changes must make assumptions about all of these factors. The empirical studies reviewed found lower reductions, between 4 and 5%, than the studies that used models, between 8 and 10%. When annual VMT is measured in the trillions, even a few percentage points make an enormous difference in determining PAYD effectiveness in reducing GHG emissions.

Data and Tools
None available.

Implementation Concerns
Agency Cost
There are no particular agency costs associated with PAYD insurance, because no additional infrastructure is required and most observers assume that, once low-mileage drivers realize they can save money, the programs will grow on their own in the marketplace.

Agency Implementation Concerns
State insurance regulators are the primary government players in legalizing PAYD insurance. Some current state policies prohibit PAYD insurance; for example, in a state requires that insurance premiums be quoted to the customer before purchasing insurance, a PAYD policy that bills the customer afterward based on miles driven would be in violation of state law. Lifting some of these restrictions may conflict with other policy goals. State DOTs and MPOs have not generally played large roles in advocating for PAYD to be legalized, but they could certainly do so.

Social Concerns
PAYD has had high consumer satisfaction where implemented (87% in one survey of a pilot program [Progressive Insurance and NCTCOG, 2007]), with most people signing up because of the opportunity to save money. While PAYD would have a positive impact on people who drive less than average—one estimate states that two-thirds of drivers would save money with PAYD insurance (Bordoff and Noel, 2008)—it is not clear what the adoption or satisfaction rate would be among drivers who drive more than average, since presumably their premiums would increase. This has so far been avoided by pricing strategies that implement PAYD as a series of discounts, under which no driver pays more, and because PAYD is of course voluntary. It seems possible that PAYD could result in a system, at least in the short-term, in which low-mileage drivers pay their fair share, while high-mileage drivers continue to be subsidized. Bordoff and Noel (2008) assume that as adoption becomes more widespread, insurance companies would be forced to raise their rates

on high-mileage drivers, resulting in a "virtuous circle" in which drivers would be compelled to drive less to keep their insurance rates low.

Depending on the technology used, some drivers may be reluctant to switch to PAYD insurance for privacy reasons (e.g., if they perceive that the insurance company is tracking where they drive). Insurance companies in other countries have implemented PAYD with technologies such as global positioning systems (GPS) and other types of on-board units that plug into the vehicle's on-board diagnostics port to record data related to speed, which can then be used to determine mileage. While all types of on-board units can be configured to provide only the number of miles driven, and not the location, drivers may not be convinced that such protections are in place.

Other Costs/Barriers

Bordoff and Noel (2008) pointed to three key barriers to adoption: the difficulty of monitoring mileage driven, state insurance regulations, and patented technology. First, various technologies exist today to meter mileage or otherwise tie insurance coverage to miles driven, although privacy concerns and expense make them difficult to adopt. One experiment with PAYD by British insurance company Norwich Union purportedly ended when the equipment cost was found to be too high relative to the program's benefits (Norwich Union axes "Pay as you Drive" Scheme, 2008). Second, many state insurance regulations (e.g., which require stating the premium cost up-front) prohibit or conflict with PAYD characteristics such as basing payment on miles driven. Third, Progressive Insurance, the only U.S. company to offer PAYD insurance with an after-market technology for metering mileage, has obtained patents that seem to make it difficult for other companies to bring similar technology to market (Bordoff and Noel, 2008).

It is possible that drivers may fraudulently try to lower their premiums by reporting lower mileage than they actually drove. A federal report on odometer fraud found that the possibility of an odometer being rolled back (tampered with to show a lower number of miles driven) is about 3.5% over the first 11 years of the vehicle's life. Annually there are about 450,000 cases of odometer fraud (NHTSA, 2002). Carfax, a private company that supplies vehicle reports, says, "Digital odometers, thought to be the answer to odometer tampering and fraud, are as easy, if not easier, to alter as their mechanical predecessors." (Carfax, 2010).

Insurance companies may experience lower premium revenues. However, if PAYD programs encourage less or safer driving, accidents could be avoided and thus companies could save on insurance payouts.

Interactions with Other Strategies
None identified.

Unique Co-benefits
- If PAYD is voluntary, it is likely to be adopted by those who drive less and thus result in reduced insurance payments.

Unique Negative Effects
- PAYD may have negative impacts for high-mileage drivers, whose premiums would ultimately rise. On the other hand, this may induce them to also drive less.

Where in Use
Currently 34 states allow some form of PAYD insurance, as do a number of countries in Europe and Asia (EDF, 2009). However, only one company in the U.S., Progressive Insurance, currently offers PAYD on a mileage basis in nearly 20 states. GM vehicle owners can apply for mileage discounts if their vehicles are equipped with OnStar, a GPS system, and Milemeter (which operates only in Texas) offers PAYD insurance based on odometer readings. There are no published figures on the number of American drivers who currently have PAYD insurance. PAYD has been more widely adopted in Europe, Asia, and South Africa.

Recommendations for Further Research
Studies of market penetration in other countries where PAYD has been adopted may exist. Such studies should be assessed for lessons learned and to give an indication of how PAYD might fare in the U.S.

References
Bordoff, J. E., and Noel, P. J. (2008). *Pay-as-you-drive auto insurance: A simple way to reduce driving-related harms and increase equity.* Washington, DC: The Brookings Institution, Discussion Paper 2008-09.

Bureau of Transportation Statistics (2009). *National Transportation Statistics, 2009.* U.S. Department of Transportation. Available online at: http://www.bts.gov/publications/national_transportation_statistics.

Cambridge Systematics, GeoStats, and MarketLine Research. (2006). *Mileage-based user fee demonstration project: Pay-as-you-drive experimental findings, final report.* Minneapolis, MN: Minnesota Department of Transportation, MN/RC – 2006-39A.

Carfax. (2010). Uncovering Odometer Fraud. http://www.carfax.com/car_buying/odometer.cfx, accessed on March 1, 2010.

Edlin, A. S. (2003). "Per-mile premiums for auto insurance." In *Economics for an imperfect world: Essays in honor of Joseph Stiglitz.* Cambridge, MA: MIT Press.

Environmental Defense Fund. (2009). Drive less, pay less for insurance. Web page at www.edf.org/page.cfm?tagID=31651, accessed December 16, 2009.

Litman, T. (2009). "Pay-As-You-Drive Pricing For Insurance Affordability," Victoria Transport Policy Institute.

National Highway Traffic Safety Administration. (2002, April). Preliminary Report: The Incidence Rate of Odometer Fraud. DOT HS 809 441, NHTSA Technical Report.

Norwich Union axes "Pay as you drive" Scheme. (2008, June 18). Motortrader.com.

Parry, I. W. H. (2005). *Is pay-as-you-drive insurance a better way to reduce gasoline than gasoline taxes?* AEA Papers and Proceedings, 95(2), 288-293.

Progressive County Mutual Insurance Company, and The North Central Texas Council of Governments. (2007). *Pay as you drive (PAYD) insurance pilot program, phase 2 final project report.*

Steininger, K., Vogl, C., and Zettl, R. (1996). *Car-sharing organizations: The size of the market segment and revealed change in mobility behavior.* Transport Policy, 3(4), 177-185.

Ridesharing and HOV Lanes

Policy: Most vehicle trips, especially commuter trips, are taken in single-occupant vehicles (SOV). The capacity of the existing roadway network could be increased if vehicle occupancy increased. Ridesharing strategies include conducting outreach programs and providing services to increase carpooling and vanpooling, and thereby reducing VMT and GHGs). Agencies can also create high-occupancy vehicle (HOV) lanes, which enable ride sharers to avoid congestion or tolls, serving as a further inducement.

Emissions Benefits and Costs: The emissions effects of ridesharing vary greatly depending on the types of policies used to encourage it and the context in which it is encouraged. Additionally, many studies report vehicle occupancy rates and mode share rather than GHGs. For these reasons, generalizations about GHG effects cannot be made. Additionally, costs to implement ridesharing programs are often bundled with other programs, so it is difficult to develop cost estimates. There is even less evidence about the effects of HOV lanes; some studies have found increases in ridesharing along HOV corridors, while others have not, and no studies were identified that assessed GHG emissions specifically. Importantly, creating new HOV lanes produces GHGs, and these life-cycle emissions must be considered in an assessment of HOV effectiveness as a GHG mitigation strategy.

Implementation Concerns: For ridesharing programs, concerns are few: they are widely implemented and well accepted given the benefits they provide to travelers. For HOV lanes, there is concern that they take away capacity from SOV driving and create more congestion. Recently there has been a trend toward high-occupancy toll (HOT) lanes, which are thought to be a more effective means of managing demand. HOV lanes may also involve construction costs if new lanes are created, which are much higher than other TDM strategy costs.

Background

In 2001, just over 85% of all trips in the U.S. were made by car, and 65% of those car trips were single-occupant vehicles (SOVs) (NHTS, 2001). One aphorism in transportation planning is that the most underutilized capacity is the three or more empty seats in every SOV. If the same number of person trips were made in fewer vehicles, the transportation system would operate more efficiently. Moving a larger number of people with the same capacity, fuel consumption, and GHGs is an obvious way to increase efficiency.

Ridesharing, of course, occurs without any policy intervention, since many people are willing to share rides for convenience, cost savings, or company. This strategy seeks to increase the amount of ridesharing, particularly for commuter trips, which are more likely to be made in SOVs than other trip types. According to various surveys reviewed in *Commuting in America*, the SOV mode share for commute trips was about 75 to 77% in the early 2000s, an increase from 65% in 1980. Most of this change resulted from a decline in ridesharing, from 19 to 12% (Pisarski, 2006).

U.S. Department of Transportation
Federal Highway Administration

Policy and Implementing Organizations

Ridesharing is generally divided into carpooling, in which ride sharers use their personal vehicles, and vanpooling, in which employers provide group transportation in larger vans and buses. Most efforts to increase carpooling and vanpooling are made at the regional level by commuter assistance organizations. In some regions, high-occupancy vehicle (HOV) lanes exist to encourage ridesharing.

While no special accommodations are needed for people to carpool, several strategies have been used by commuter assistance organizations to increase carpooling. One strategy is to provide rideshare matching services, which allow prospective ride sharers to find others who work and live near them. Second, many firms provide "dynamic ridesharing," which makes quick matches online for one-time rides (as opposed to conventional matching systems in which both ride sharers are interested in ridesharing for an extended period of time). Third, employers can encourage carpooling through preferred parking, cheaper parking rates for carpoolers and others, and commute assistance organizations often encourage employers to adopt such policies.

In a few areas, "casual carpooling" has become possible. That is, in order to gain access to HOV lanes and/or avoid tolls within these regions, solo drivers pick up passengers who wait at designated pick-up sites, often at park-and-rides or along transit routes, and bring them to designated drop-off points, generally in central business districts or other high-employment areas. Drivers and passengers who participate in casual carpooling generally agree to a few rules, which tend to be self-enforced, and safety has not proven to be a major issue with these informal programs. Local governments may assist such programs by installing signage, though they can operate independently.

A key difference between vanpools and carpools is that vanpools generally charge riders a fee to cover operating expenses, and federal law also provides a tax credit for vanpoolers (but not carpoolers). Commuter assistance organizations also promote vanpooling, for example by providing technical assistance (for example, working with an employer to set up a vanpooling program), by operating vanpools, or by providing direct subsidies. Some organizations, such as transportation management associations,[51] also promote or operate vanpools, and there are several commercial vanpool providers.

HOV lanes enable ride sharers to avoid congestion—and in some cases, tolls—by designating specified lanes off-limits to SOVs. The number of occupants required in order to use HOV lanes varies by region; sometimes two people are required, sometimes three. The time of day that the

---

[51] Transportation Management Associations (TMA) are generally not-for-profit organizations that promote alternatives to SOV commuting in specific areas, such as an office park or corridor.

U.S. Department of Transportation
Federal Highway Administration

lanes are restricted can vary as well; some operate during peak hours only, others 24 hours a day. HOV lanes can be converted from traditional lanes, or built as new lanes.

Effects

Target Group

Like other TDM policies, ridesharing strategies target both employers and employees. HOV lanes can be used by all travelers, although many HOV lanes are in effect only during commute hours.

GHG Effects

While ridesharing trends in the U.S. are well documented, there is little formal evaluation of the effectiveness of ridesharing promotion. One reason it is difficult to measure the impact of either ridesharing programs or the presence of HOV lanes on carpooling is that a number of other factors affect drivers' willingness to form carpools. A review of the literature on carpool formation in Parkany (1998) noted that factors such as cost of SOV driving, distance to work, education levels, whether the driver is a professional, the number of employers in an area, gender, and household size all play a role in decisions to carpool.

Importantly, people are more likely to rideshare for trips other than commuting, since much natural ridesharing occurs between family members. In 2001, the average vehicle occupancy for work trips was 1.13, while for social trips it was 2.03; the average for all trips was 1.63 (Hu and Reuscher, 2001).

The studies cited below largely rely on commuter surveys, not on observed behavior. The few reports that have been published on casual carpooling and dynamic ridesharing do not examine their effectiveness in reducing SOV driving. There have been some evaluations of HOV lane effectiveness in the U.S.

The Metropolitan Washington (D.C.) Council of Governments (MWCOG), which does much ridesharing assessment, relies on commuter surveys to assess the programs' impacts. For FY 2003 to 2005, their integrated ridesharing program (consisting of online ridematching as well as stand-alone interactive kiosks located throughout the region) reduced vehicle trips by 5,600 and reduced 146,000 VMT per day (LDA Consulting et al., 2005). Assuming an average fleet fuel economy of 20.7 mpg[52], this means a reduction of 62 $MTCO_2$ per day. In the following evaluation period (FY 2006 to 2008), upgrades to rideshare software that supports the commuter operations center were evaluated separately[53] and were found to reduce daily trips by 4,500 and VMT by 84,000, and

---

[52] The average fuel economy for the overall fleet of cars and light trucks in the study was calculated from data from the Bureau of Transportation Statistics as noted in Table 3.1.

[53] The FY 2003-2005 and FY 2006-2008 evaluations looked at different activities, so they are not directly comparable.

annual MTCO$_2$ by 15,100[54] (LDA Consulting et al., 2009). Percentage decreases from the baseline were not provided. The FY 2008 overall evaluation for all D.C.-area commuter programs combined found an aggregate reduction of 264,500 MTCO$_2$ per year[55] (NCRTPB, 2009).[56]

In Atlanta, an evaluation of carpooling and ridesharing found a total daily trip reduction of 8,170 (5,500 attributed to carpooling and 2,670 to vanpooling) and net daily VMT reductions of 218,000 (127,000 to carpooling and 91,000 to vanpooling) (CTE, 2002). A later evaluation of four TDM measures, three of which were related to ridesharing (rideshare placement, vanpooling, and cash incentives to switch from SOV driving) found their combined impacts to be a daily reduction of 41,000 vehicle trips and 885,000 VMT. For both studies, percentage decreases from the baseline were not provided, and the study did not assess GHG reductions (CTE et al., 2004). For illustration purposes, if one (reasonably) assumes that the displaced trips took place equally in cars or light-duty trucks, which together have an average fuel economy of 20.7 mpg[57], the daily reductions from the 885,000 VMT reduced is about 380 MTCO$_2$.

The state of Washington has estimated that its commute trip reduction program, which included a variety of TDM initiatives, reduces VMT by 170 million per year, or 680,800 per day, and emissions by 85,700 MTCO$_2$ per year, or 342 MTCO$_2$ per day (CTR Interim Report to the State Legislature, 2007).

In the 1980s, the Los Angeles region adopted several mandatory developer- and employer-based travel demand management programs. Developers in certain areas had to ensure that new developments reduce the number of SOV trips, and all employers with more than 100 employees had to reduce the number of SOV trips to their worksites to a specified amount. A study of one developer-based program found that carpooling was twice as high at buildings covered by the ordinance than at buildings that were not (7.4 vs. 3.5%) (Blankson and Wachs, 1990). Evaluations of the employer-based programs found that after one year, the average vehicle ridership (AVR) during the morning peak increased from 1.213 to 1.246, and for employers who participated for two years, it increased from 1.258 to 1.304 (Giuliano et al., 1993). Neither study reported trip distances, so it is not possible to estimate emission reductions.

---

[54] Originally reported as 16,669 short tons.

[55] Originally reported as 291,608 short tons.

[56] The commuter programs include support for employer transit benefits, but if employers use transit benefits without COG assistance, those effects are not counted directly in the study. Programs like parking cash-out are not included.

[57] The average fuel economy for the overall fleet of cars and light trucks was calculated from data from the Bureau of Transportation Statistics as noted in Table X.

U.S. Department of Transportation
Federal Highway Administration

In a study of casual carpooling, the two existing casual carpool systems in the San Francisco Bay Area and northern Virginia—which account for an estimated 3,000 and 3,500 carpools per weekday—were found to save about 3 million gallons of gasoline per year. The report estimated that a group of 150 commuters who switched from SOV commuting to casual carpooling would save about 52,000 gallons of gasoline per year, roughly the same as an express bus service. This was based on assumptions of 12-mile commutes and HOV lanes with higher traffic speeds than the general-purpose lanes (Dorinson et al., 2009). This amounts to approximately 460 MTCO$_2$.

While HOV lanes may be a factor in individuals' decisions to rideshare, the extent of this is unknown and depends on many factors. Our review found that HOVs have a mixed record of promoting rideshare formation and that relatively little information on emissions impacts is available. In terms of emissions, older studies (from the 1970s) estimated reductions in fuel consumption ranging from 7-10% to up to 26% (Turnbull et al., 2006). A California study also found that HOV lane emissions rates (for criteria pollutants) were about half of the adjacent free lanes, but the study did not consider the extent to which HOV lanes may have contributed to congestion and emissions in the free lanes. The study did not assess GHG emissions (Parsons Brinckerhoff Quade and Douglas et al., 2002). A modeled study of returning HOV lanes to general purpose lanes in Minneapolis found a *savings* in fuel consumption of 4,000 gallons per day because of increased speeds throughout the region (Cambridge Systematics and URS, 2002). A recent overview of the literature on HOV lanes and emissions concluded that there is a "lack of in-depth information on the air quality, energy, and other related environmental impacts of HOV facilities" (Turnbull et al., 2006).

One report on California, which has more HOV lanes than any other state, found that a number of carpoolers in the San Francisco Bay Area cited the HOV lane as a factor in their decision to carpool. Survey data from the rest of the state was not available. In southern California, HOVs saw increases of 25 to 35% in peak period carpools compared to highways without carpools (Long, 2000). An evaluation of the HOV lanes in southern California found that about half of all carpools using the HOV lanes were formed in response to the HOV lane, and that average vehicle occupancies have increased on the facilities with HOV lanes compared to two control routes (Parsons Brinckerhoff Quade and Douglas et al., 2002).

A study in Dallas found that AM peak hour carpools at least doubled on all four HOV segments, and that average vehicle occupancy increased by 8-12%, while a control route without HOV experienced a 2% decrease in vehicle occupancy over the same time period. The HOV lanes also carried more persons per lane than the free lanes, a key measure of HOV lane efficiency (Skowronek et al., 1999).

Vancouver, Washington added a new HOV lane to an existing highway, in part to improve travel time reliability for carpools, vanpools, and bus transit. A study of this lane found increases in bus transit use of 18% in the first two years after HOV lanes were implemented. While baseline figures were not available to examine the growth in carpooling, the number of people using the HOV lane was nearly double the number using the free lanes before the HOV opened, suggesting that some

carpools and vanpools must have been newly formed. Emissions reductions were not evaluated (Parsons Brinckerhoff Quade and Douglas, 2004).

In Oregon, an evaluation found that vehicle occupancy increased from 1.37 to 1.39, suggesting that carpool rates increased from 37 to 39% (quoted in Martin et al., 2005). A study in Utah found that during the AM peak, the HOV lanes carried fewer people than the average free lane (1,267 vs. 1,549), while in the PM peak they carried about 8% more people (1,700 vs. 1,568). Pre-HOV baseline figures were not reported, so the study could not measure whether total vehicle occupancy increased (Martin et al., 2005).

Finally, if new lanes are being built specifically for HOV use, the emissions from construction may be significant and must be taken into account in order to know the true effect of HOV lanes as a GHG mitigation strategy.

Estimated Cost per Metric Ton of CO_2 Reduction

Without more detailed data of the impacts of ridesharing programs on behavior changes, this is very difficult to estimate. The Washington, D.C. region estimated a cost per CO_2 ton reduced of $15, but this included all commuter assistance and was not specific to ridesharing. The report noted that, "The Commuter Connections Program is generally regarded as among the most effective commuter assistance programs in the nation in terms of reductions effected in vehicle trips and vehicle miles of travel" (NCRTPB, 2009). If this is correct, and the authors have found no evidence to the contrary, other regions would have higher costs per ton. With so little information available on the GHG impacts of HOV lanes, it is impossible to make a reliable calculation for these strategies.

Key Assumptions and Uncertainties

The largest uncertainty in estimating effects is the degree to which SOV drivers respond to incentives to rideshare and to the availability of HOV lanes. As noted above, many factors can influence these decisions, so it is very difficult to assign impacts to specific TDM measures. Although the regional evaluations cited in the preceding section assigned such impacts, they are based on commuter surveys and not observed behavior, and make assumptions about the extent to which self-reported behaviors reflect actual changes in behavior.

Studies may also make assumptions about unintended effects of ridesharing. For example, although each ridesharing trip may remove one or more vehicles from the road, the vehicle that is being used is likely to travel farther to pick up or drop off each passenger. Side trips may increase for ride sharers if they are no longer able to combine activities like picking up groceries on a commute trip. These effects seem likely to be small in comparison to the VMT reductions, but many assessments disregard the effects entirely.

Where new HOV lanes are created, the GHG emissions from HOV lane construction, operations, and maintenance may be unknown but could reduce or even negate the benefits of roundabouts. Where HOV lanes are converted from traditional lanes, they may increase congestion and emissions in the free lanes. As cited earlier, there is a "lack of in-depth information on the air quality, energy, and other related environmental impacts of HOV facilities" (Turnbull et al., 2006).

Data and Tools

The evaluation methodology of ridesharing and other TDM programs developed by the Washington Metropolitan Council of Governments is one of the most sophisticated in use in the U.S. The techniques are described in LDA Consulting et al. (2007).

Implementation Concerns

Agency Cost

As noted elsewhere, ridesharing is not often treated separately from other commuter assistance programs, and the same is true for ridesharing budgets. In Washington, D.C., the annual budget for the Commuter Connections program is $5.2 million, including staff time, operating the ridematching database, and marketing (NCRTPB, 2009). In Washington State, costs for the two-year period from 2007 to 2009 were $7.3 million, or an average of $3.65 million per year; again, this includes all components of commuter assistance, not just ridesharing (WSDOT, 2009). These are large programs; many regions presumably operate with far smaller budgets. As most regions currently operate commuter assistance programs, one cannot estimate start-up costs here.

Costs for HOV lanes vary since some have been converted from existing capacity, and others have been built as new construction. Adding one lane mile to an urban highway is estimated to cost roughly $10 to $15 million (FHWA, 2008).

Agency Implementation Concerns

Ridesharing efforts need to be sustained over time. As carpools dissolve, people and worksites move, and new employees and employers enter a region, rideshare matching efforts and more general education about travel demand management must be ongoing.

Social Concerns

Ridesharing on a voluntary basis is already a widely accepted strategy. While several areas have passed mandatory TDM ordinances, these tend to be more controversial. Los Angeles had fairly stringent requirements in place that were eventually softened due to pressure from the business community.

HOV lanes have met with controversy as well, sometimes because they are perceived as taking capacity away from SOV drivers in congested free lanes, and sometimes on the environmental grounds that in freeing capacity they induce more travel demand for driving (Turnbull et al., 2006).

Other Costs/Barriers

Barriers to increased ridesharing include difficulties in finding rideshare partners, lack of schedule flexibility, and low commute costs. Some of the difficulty in finding partners can be solved with rideshare matching services, while others are linked to decentralized workplaces (since the odds of finding a good rideshare partner, or a vanpool, presumably rise with a higher residential density and higher density of jobs, living and working in low-density locations can make it more difficult).

Vanpools can also have problems since vanpools are generally paid services and must have a certain number of riders to remain viable. This is less of a structural problem and stems from the need to do some continuous marketing and outreach to identify new riders when previous riders drop out for whatever reason.

Interactions with Other Strategies

- Ridesharing complements other employer-based TDM strategies, such as vanpool benefits (a type of financial incentive under which employees can receive employer-paid benefits or use pre-tax income to pay for vanpool expenses), employee rewards for non-SOV commuting, or parking management and pricing.

- Like HOV lanes, HOT lanes may also encourage ridesharing by allowing ride sharers to use designated lanes without penalty.

- Ridesharing may be less effective when implemented along with strategies that promote other modes, since persons who start ridesharing may have previously used transit, walked, or bicycled. The benefits of ridesharing are greatest when new ride sharers previously drove alone.

Unique Co-benefits

- Reduced cost for drivers.

Unique Negative Effects

None identified.

Where in Use

Most metropolitan regions have a commuter assistance program whose function is to decrease SOV commuting in a region. These programs generally work with employers not only to encourage employees not to drive alone to work, but also to provide services to employees such as rideshare matching. Many also do general outreach through media campaigns and special promotions ("try transit" or "bike to work week") to raise the public's awareness of commuting options. In three regions—Seattle, Southern California, and Tucson—these employer programs are mandatory; in other areas they are voluntary. Casual carpooling takes place in the San Francisco Bay Area; Washington, D.C.; Houston; and Pittsburgh (Kelley, 2007).

HOV lanes of various types (full-day vs. only certain hours, reversible vs. permanent, etc.) have been built in 25 states as of 2007 (FHWA, 2010).

Recommendations for Further Research

As noted above, relatively little is known about the systemic impact of promoting ridesharing, and even less about the differences in particular means (general outreach and ridematching vs.

employer-based incentives, static vs. dynamic ridematching, and carpooling vs. vanpooling). It is also important to study further the long-term ridesharing retention rates, since job, schedule, and residential changes mean that people frequently return to SOV driving if ridesharing is no longer convenient. It is additionally difficult to draw conclusions about the effectiveness of such programs across different regions, especially when most of the research on these programs is from large regions.

With regard to HOV lanes, more before-and-after data could be collected and analyzed to determine their effectiveness in carpool formation and whether they move more vehicle occupants per lane per hour than conventional lanes. It might also be useful to conduct research on how effective they have been in effecting a market shift to lower-emissions vehicles; anecdotally, those HOV lanes that exempt hybrid or electric vehicles have become more crowded as those vehicles gain in popularity (Ginsberg, 2005).

References

Blankson, C. and M. Wachs. (1990, January). Preliminary Evaluation of the Coastal Transportation Corridor Ordinance in Los Angeles. Transportation Research Record.

Cambridge Systematics and URS (2002, February). Twin Cities HOV Study, Volume I, Final Report, Minnesota Department of Transportation.

Center For Transportation and the Environment (CTE). (FY2002). *Evaluation of the effectiveness of programs contained in the "framework for cooperation to reduce traffic congestion and improve air quality", phase three, FY2002 Atlanta TDM framework final report*: Georgia Department of Transportation.

Center For Transportation and the Environment, Georgia Department of Transportation, Federal Highway Administration, CIC Research Inc., Earthmatters Inc., ESTC, and LDA Consulting. (2004). *Voluntary mobile emission source program (VMEP) state implementation plan (SIP) assessment, 2004 VMEP assessment.*

Dorinson, D., D. Gay, P. Minett, and S. Shaheen. (2009). Flexible Carpooling: Exploratory Study. Davis, Institute of Transportation Studies, University of California at Davis.

Federal Highway Administration. (2010). HOV Clearinghouse, available at http://arcims.urscorp.com/HOV/frmfac.asp. Accessed on February 16, 2010.

Federal Highway Administration. (2008). Congestion Pricing A Primer: Overview. FHWAHOP-08-039. Available at http://ops.fhwa.dot.gov/publications/fhwahop08039/cp_prim1_00.htm. Accessed on September 6, 2010.

Ginsberg, Steven. (2005). Hybrids Could Lose HOV Perk Early; Va. Offers Options Aimed at Restricting Usage of I-95/395 Lanes. *Washington Post*, March 2.

Giuliano, G., K. Hwang, and M. Wachs. (1993). "Employee Trip Reduction in Southern California: First Year Results. Transportation Research vol. 27a, no. 2.

Hu, P. S., and Reuscher, T. R. (2004). *Summary of travel trends: 2001 National Household Travel Survey.* Washington, DC: U. S. Department of Transportation, Federal Highway Administration.

Kelley, K. L. (2007). Casual carpooling—enhanced. Journal of Public Transportation, 10(4), 119-130.

LDA Consulting, CIC Research Inc., ESTC, and Center for Urban Transportation Research. (2009).

Transportation emission reduction measure (TERM) analysis report, FY 2006-2008. Washington, DC: National Capital Region Transportation Planning Board Commuter Connections Program.

LDA Consulting, CIC Research Inc., ESTC, Shirazi, E., and Collier, C. (2005). Transportation emission reduction measure (term) analysis report, FY 2003-2005. Washington, DC: Metropolitan Washington Council of Governments Commuter Connections Program.

Long, R. (2000). *HOV lanes in California: Are they achieving their goals?* Sacramento: Legislative Analyst's Office.

Martin, P. T., Lahon, D., and Stevanovic, A. (2005). Review of the effectiveness of the high occupancy vehicle (HOV) lanes extension. Salt Lake City: Department of Civil and Environmental Engineering, University of Utah.

National Capital Region Transportation Planning Board (NCRTPB). (2009). Work program for the commuter connections program for the greater Washington metropolitan region, fiscal year 2010. Washington, DC: Metropolitan Washington Council Of Governments.

National Household Transportation Survey. (2001). Data from online analysis tool, accessed December 16, 2009.

Parkany, E. (1998). Can HOT Lanes Encourage Carpooling? A Case Study of Carpooling Behavior on the 91 Express Lanes, Institute of Transportation Studies, University of California at Irvine.

Parsons Brinckerhoff Quade and Douglas Inc., Kaku Associates Inc., Texas Transportation Institute, Strategic Consulting and Research, and HS Public Affairs. (2002). *HOV performance program: Evaluation report,* Los Angeles County Metropolitan Transportation Authority.

Parsons Brinckerhoff Quade and Douglas Inc. (2004). Vancouver HOV lane pilot project, evaluation report #6.

Pisarski, A. E. (2006). *Commuting in America III: The third national report on commuting patterns and trends.* Washington, DC: Transportation Research Board, NCHRP Report 550/TCRP Report 110.

Skowronek, D. A., Ranft, S. E., and Slack, J. D. (1999). Project monitors HOV lane operations Collage Station: Texas Transportation Institute, Project Summary Report 3942-S, Project 7-3942: Investigation of HOV Lane Implementation and Operational Issues.

Turnbull, K. F., Levinson, H. S., Pratt, R. H., John E. (Jay) Evans, I., and Bhatt, K. U. (2006). HOV facilities: Traveler response to transportation system changes. Washington, DC: Transportation Research Board, TCRP Report 95, Chapter 2.

Washington State Commute Trip Reduction Board. (2007). CTR interim report to the Washington state legislature.

Washington State Department of Transportation (WSDOT). (2009). Funding and structure (web site), www.wsdot.wa.gov/TDM/CTR/funding.htm, accessed on December 17, 2009.

Transit Incentives

Policy: While most cities, even small ones, have some type of transit service, the out-of-pocket fares and inconvenience of riding transit may result in low ridership. One way to encourage people to switch from driving to transit is to make transit cheaper for riders. Federal law now contains tax incentives that allow employers to reduce employees' transit fares. Transit agencies have also adopted a variety of special programs to decrease riders' costs. Together, these can reduce GHG emissions if new riders switch from driving alone.

Emissions Benefits and Costs: The effect of employers' decisions to offer transit benefits is unknown, and where benefits are offered, the effects on ridership and transit mode share can vary greatly, from 0 to 17 percentage points. The ultimate effect on GHG emissions is unknown, but could be large in some cases. Transit agency promotions to reduce fares have not been systematically studied. The administrative costs for transit agencies to facilitate employer provision of transit incentives can vary significantly depending on the type of program. Studies report costs ranging from $100,000 to $500,000 annually, which principally include marketing and staff costs. Transit incentive programs also affect transit revenues: employer-based programs tend to increase revenues while fare reduction programs can decrease revenues.

Implementation Concerns: Barriers to implementing employer-based transit incentives are generally low because the programs are voluntary for consumers and may be voluntary for employers. While they include costs to employers, these benefits typically become part of an employer's benefits package.

Background

Transit is available in many regions, albeit with different types and amounts of service and different ridership levels. To the extent that new riders can be accommodated with existing capacity, increased transit ridership can reduce GHG emissions, provided that the new transit trips replace vehicle trips, particularly SOV trips. Many regions try to increase the use of transit by generating more demand, generally by using incentives that reduce riders' costs. These incentives are most commonly provided through employer-based "transit benefit" programs, but they can also be provided with fare discounts or free ride programs for all transit users. (Increases in transit services to boost ridership are discussed separately in the discussion on Transit Improvements.)

Employer-based transit benefits are possible because of a provision of the U.S. tax code[58] that allows employers to provide direct or indirect assistance toward employees' transit fares. Until the early 1990s, employers were allowed to provide free parking as an untaxed benefit to their employees, but any assistance above $15 to ride transit would be taxed. To correct this imbalance,

---

[58] Title 26 USC Section 132(f) – "Qualified Transportation Fringe"

U.S. Department of Transportation
Federal Highway Administration

the federal government re-defined "qualified transportation fringe benefit" in 1992 to allow employers to provide transit and vanpool benefits to employees tax-free up to certain levels. The provision also requires that employers provide transit passes and vouchers in regions where they are available, instead of paying the employees directly.

Transit benefits can be provided by employers in two ways: they can give an employee a specific amount in transit costs as a direct subsidy, or they can allow an employee to purchase transit passes with pre-tax income (similar to a health flexible spending account). They can also combine these two options so that an employee can receive a subsidy and set aside additional pre-tax income.

Policy and Implementing Organizations

Two types of policies as transit incentives are identified below:

The first policy is transit benefits. As with other employer-based tax provisions, such as 401(k) plans, the employer must offer the benefit before an employee can utilize it. Therefore, one policy is for a public agency to market transit benefits to employers, encouraging them to provide their employees with transit benefits. The agencies that typically promote transit benefits to employers include commuter assistance organizations, transit agencies, and transportation management associations. In some areas, employers having a certain number of employees are subject to mandatory commuter trip reduction programs, in which employers must take actions to try to reduce the number of SOV commute trips. Transit benefits can be one way of fulfilling this mandate.

The second policy is for transit agencies to incentivize transit use by reducing the cost to riders. For example, transit agencies can implement universal pass programs that offer deeply discounted fares to employers on behalf of their employees, provided employers purchase passes for all or a certain portion of their workforce. Other incentives such as discounted or free fares can be directly offered to all riders, either on a permanent or promotional basis.

It is possible to use these policies together or separately. For example, Washington, D.C. employers make heavy use of employer-based transit benefits programs, but their employees do not receive any discounts from the regular rail fare. Universal pass programs, such as Eco Pass in Denver, are offered to employers at deep discounts, who then provide them as a transit benefit to employees. Portland's "Fareless Square" provides free rail transit to all riders within an area of downtown, regardless of their employment situation.

Effects

Target Group

Like other TDM strategies, employer-based transit benefit strategies target both employers and employees because employees cannot take advantage of the tax benefits unless employers implement transit benefit programs. More general fare incentives are widely aimed at existing and potential transit riders.

GHG Effects

There are no known studies that measure how effective public sector efforts are in persuading employers to offer transit benefits. Further, no studies develop national models or estimates. However, there is some literature on the effects of employer-based transit benefit programs on transit ridership at individual workplaces and in regions. There is also some research on the effectiveness of fare policies on transit ridership, and it suggests that external factors are more influential than fares (See the review of Transit Improvements).

Research suggests that the effect of transit benefit availability on ridership and mode share can vary significantly. A review based on survey data from 13 cities (ICF and CUTR, 2005) found that, where employers implemented transit benefits voluntarily (that is, they were not required to offer benefits), transit ridership—the number of employees riding transit on a given day—at those worksites increased by at least 10% (results were reported for cities as a whole, not for individual worksites). Gains were made both from new riders and from increased transit use from existing transit riders. While the review included various pass types, there was not enough data to determine whether certain pass types were more effective than others.

Specifically, increases in ridership at worksites ranged from 10% to over 150%, with about half of all surveys finding increases between 10% and 40%. Perhaps a more important indicator than gains in ridership is the increase in transit mode share before and after benefits were offered. In seven of the thirteen cities for which this information was available, the increases in transit mode share ranged from about 2 to 17 percentage points. The starting mode shares ranged from under 10 to over 40%.

In the ICF and CUTR review, transit benefits that were implemented in response to a mandatory commuter trip reduction (CTR) program (which *requires* employers to offer transit benefits or other incentives not to drive alone) had little and sometimes no effect on transit ridership. Interviews with CTR staff revealed that many employers implemented transit benefits to meet regional requirements, since they would receive credit for an employer program, even if few or no employees used the benefit (ICF and CUTR, 2005). The mandatory programs that this report addressed were in southern California, Washington State, and Tucson, Arizona. These have been less effective than the mandatory program enacted under Regulation XV in Los Angeles, which imposed much more stringent requirements on worksites to lower their average vehicle occupancy. However, pressure from the business community brought about its repeal (Sorensen et al., 2008).

ICF and CUTR (2005) did not estimate CO_2 reductions from the increases in transit ridership, and this would vary significantly depending on the emissions from a trip on the original mode (which depend on the vehicle type, distance, system efficiency, etc.) and the emissions from a trip via transit. Nevertheless, the report does note that in seven of the twelve cities, 90-100% of new transit commuters were previous single occupancy vehicle (SOV) commuters. In seven of the 12 cities that had such data, SOV ridership at the surveyed employers declined between 1 and 15 percentage points.

U.S. Department of Transportation
Federal Highway Administration

The ICF and CUTR report also looked at the regional impact of transit benefits, but data was available in only three regions. The impact was largest in Washington, D.C., where a change that required *federal* employers to provide transit benefits was estimated to have resulted in a 29% increase in overall ridership, or 60,000 new riders per day. In Denver, the Eco Pass program (a universal pass program available to employers) was estimated to have brought in 6,000 new riders. In San Jose, the report estimated that the Eco Pass program might have brought in 16,000 new riders; however, the estimate was based on a small number of worksites, so the accuracy of this estimate is unclear. Unfortunately, because of the paucity of data at the regional level, and the many determinants of transit ridership, this sourcebook cannot provide even a range of potential regional impacts.

For those regions where data was available, one can approximate the emissions reductions achieved by programs by considering the extent to which the new trips replace SOV trips. Denver and San Jose were among the cities in which over 90% of new transit riders were previously SOV drivers; in Washington, 60% were SOV drivers. Assuming an average fleet fuel economy of 20.7 mpg[59] and a 12.1-mile average one-way commute (Hu and Reuscher, 2001), this yields reductions of roughly 55 $MTCO_2$ per day in Denver, 150 in San Jose, and 375 in Washington.[60]

Transit incentives in the form of universal pass programs have generally shown positive results in increasing transit ridership in one particular case: universities. A number of universities have implemented such programs for their students, faculty, and/or staff. At the University of Washington, one of the first campuses to institute such a program, student transit mode share rose from 21 to 35%, and faculty/staff from 21 to 28%; SOV shares fell in both groups. At the University of California at Berkeley, student transit mode share rose from 5.6 to 14.1% (Nuworsoo, 2005). At UCLA, faculty/staff transit mode share for people living within the bus service area grew from 9 to 20%, and for students, from 17 to 24%. While both groups reduced their SOV use, there were declines in other modes as well, indicating that some people switched from carpooling, vanpooling, and bicycling (Brown et al., 2003).

BART, a heavy rail system in the San Francisco Bay Area, estimated GHG reductions for a variety of programs that it could implement with regard to transit fares and discounts. A "kids ride free" program, allowing free Saturday travel when accompanied by an adult, would reduce an estimated 15,000 $MTCO_2$ annually. Unlimited ride passes, which would allow unlimited rides for a period of time within certain zones, were estimated to reduce 85,000 $MTCO_2$ annually. Universal passes were

---

[59] The average fuel economy for the overall fleet of cars and light trucks in the US in 2008 was 20.7 mpg, as noted in Table 3.1.

[60] This only accounts for 60% of the effect of the programs in DC—because the other 40% of riders were not driving alone before they switched to transit. Thus, it is assumed 60% of new riders eliminate two 12.1-mile SOV trips. For all cities, does not take into account the extent to which transit riders first drive to stations.

U.S. Department of Transportation
Federal Highway Administration

estimated to reduce 148,000 MTCO$_2$ per year, provided they were given to 10% of all adults in BART's service area (Nelson\Nygaard Consulting Associates, 2008).

Estimated Cost per Metric Ton of CO$_2$ Reduction

It is not possible to estimate this at the national level with available data for several reasons:

- the effects of marketing campaigns on transit benefit program implementation are unknown;
- the effects of transit benefit programs on transit ridership, and the resulting CO$_2$ reductions, vary greatly from region to region;
- transit benefits and fare reduction programs can be implemented in many ways; and
- transit benefits and fare reduction programs also have uncertain effects on transit revenues.

In sum, neither the GHG reductions, nor the overall costs, can be effectively generalized.

BART estimated costs per metric tons of GHG reduced. The "kids ride free" program would cost from -$10 to $185 per MTCO$_2$ reduced (-$10 because under one scenario the program would actually generate revenue from new adult riders). The unlimited ride passes would cost about $120 per MTCO$_2$ reduced, while the universal passes would be $150 per MTCO$_2$.

Key Assumptions and Uncertainties

A number of factors affect whether providing transit benefits increases the number of riders at a workplace. Workplaces in auto-oriented, suburban locations, with little transit service, low benefits levels, other competing TDM programs, and lots of free parking will probably see a relatively small absolute increase in transit use, even when benefits are provided. Unfortunately, the data are not robust enough to determine the impacts of specific factors. In addition, it can be difficult to try to make comparisons across cities based on the level of transit supply. Transit agencies define their service areas differently, and multiple operators often serve one region, making it difficult to construct an objective measure of the level of transit availability in a region. It is even more difficult to make these comparisons across neighborhoods.

The effect of increased ridership on CO$_2$ reductions, in turn, depends largely on the previous mode of transportation and, specifically, the percent of new riders who switched from SOV to transit. This figure can also vary widely (ICF and CUTR, 2005).

Data and Tools

None identified.

Implementation Concerns

Agency Cost

Agency costs to administer transit benefits programs (that is, the cost of the marketing, outreach, and fulfillment) are generally in the range of $100,000 to $500,000 annually, a fraction of the revenues that these programs produce (for most agencies, the administrative cost is just a few percent of the revenues) (ICF and CUTR, 2005).

A potential concern with discounted fare programs is the potential loss of revenue to the transit agency. While most transit agencies obtain over half of their operating revenues from sources other than the fare box, an agency facing a deficit may be considering raising fares, rather than reducing them. While no analyses of this issue appear to exist, anecdotally it seems some transit agencies are looking to tighten their discount programs for financial reasons (Sun Media, 2009; Grynbaum, 2009). This is probably less of an issue for those transit agencies that do not offer discounts to employers, in which case employees use their transit benefits to pay the full fare.

Agency Implementation Concerns

As transit benefits can be implemented by multiple agencies (generally, transit agencies, MPOs, and TMAs), some effort may be required to ensure that the most effective institutional structure is in place for each particular region. Agencies should not either leave gaps with regard to their target markets, or spend undue effort on overlapping initiatives. Transit benefits can also be implemented in many ways, so developing programs appropriate to the transit service and the audience may be challenging. Finally, setting appropriate prices and, if needed, determining how revenues will be divided among multiple agencies are important issues.

Social Concerns

Transit benefits are largely acceptable and are already fairly widespread, although the level of use varies from region to region. Use of transit benefits, even if the employer is required to purchase a pass, is voluntary. Transit fare programs, since they decrease riders' costs, are generally well-accepted.

Other Costs/Barriers

When employers provide direct subsidies to employees, they can incur fairly large total costs, depending on the number of participating employees and the transit fare structure in the region. Employers generally regard these non-taxable costs as part of a benefits package. Where employees set aside pre-tax money, there are small tax savings (perhaps 5% of the amount) to the employer, since those monies are exempt from payroll taxes. Since programs are voluntary, employers generally weigh these costs against other employee benefits when determining whether to implement them.

Interactions with other Strategies

- While transit benefits can be implemented on their own, transit improvement strategies may create a wider pool of potential riders.

- Road pricing, parking management, and paid parking may contribute to transit demand. Drivers for whom parking has become scarcer or driving more expensive can be compensated with transit benefits.

- Other employer-based TDM programs may attract potential riders away from transit. While this would affect the number of employees who switch to transit, if the other mode has lower CO2 emissions than SOV driving, the overall impact would not be negative.

Unique Co-benefits

- Transit incentives can result in increased ridership and revenues to transit agencies, especially when riders using transit benefits pay the full fare or when large employers or universities subsidize transit use by their employees and students.

Unique Negative Effects

- If discounts are too steep, fare reductions can result in reduced revenues to transit agencies, which may affect their ability to offer other services.

- Additionally, if transit benefits result in increased ridership during peak hours, when service is already at capacity, it can require the agency to add new service. This may be a financial burden to the agency, especially since transit fares do not cover the cost of providing service.

Where in Use

Most U.S. transit agencies that serve mid- to large-size regions have an employer-based transit pass program, which is marketed to employers either by the transit agency or other agencies. Almost all transit agencies provide some type of discounted fare, whether for students, the elderly, or disabled persons. The use of programs such as Portland's Fareless Square is much less common. As of the late 1990s, about 35 universities had universal pass programs (Brown et al., 2001)

Recommendations for Further Research

While transit agencies and commuter assistance organizations may have data on the effects of transit benefits on workplace commute modes, much of it is unpublished. An effort to collect and analyze this data would offer a better understanding of the impacts of transit benefits programs.

References

Brown, J., Hess, D. B., and Shoup, D. (2001). Unlimited Access. Transportation, 28, 233-267.

Brown, J., Hess, D. B., and Shoup, D. (2003). BruinGO: An Evaluation University of California Transportation Center, University of California at Berkeley.

Bureau of Transportation Statistics (2009). *National Transportation Statistics, 2009.* U.S. Department of Transportation. Available online at: http://www.bts.gov/publications/national_transportation_statistics.

Grynbaum, M. (2009). "Seeing Political Pressure in Proposal to Cut Student Transit Fare Subsidy." *New York Times*, December 16.

Hu, P. S., and Reuscher, T. R. (2004). Summary of travel trends: 2001 National Household Travel Survey. Washington, DC: U. S. Department of Transportation, Federal Highway Administration.

ICF Consulting, and Center for Urban Transportation Research (CUTR). (2005). *Analyzing the effectiveness of commuter benefits programs.* Washington, DC: Transportation Research Board, TCRP Report 107.

Nelson\Nygaard Consulting Associates. (2008). BART Actions to Reduce Greenhouse Gas Emissions: A Cost-Effectiveness Analysis: San Francisco Bay Area Rapid Transit District.

Nuworsoo, C. (2005). Discounting Transit Passes. Access(26), 22-27.

Sorensen, P., M. Wachs, E. Y. Min, A. Kofner, L. Ecola, M. Hanson, A. Yoh, T. Light, and J. Griffin (2008). Moving Los Angeles: Short-Term Policy Options for Improving Transportation. Santa Monica, CA, RAND Corporation.

Sun Media. (2009). "Transit fare hike is in the cards; TTC deficit blamed on Metropass." *Toronto Sun,* September 24.

Transit Improvements

Policy: Many TDM strategies seek to reduce single-occupant vehicle trips. One mode that may absorb many of these trips is transit, especially for longer trips where walking or bicycling are not feasible. However, in many regions, transit is not a viable alternative to driving because the areas are not served by transit, the service frequencies are too low, or transit is not viewed as desirable. Transit improvements are aimed at increasing the potential for transit to absorb higher shares of trips by either creating new routes, increasing service frequencies, or increasing the comfort of transit to make it more attractive to potential riders.

Emissions Benefits and Costs: Emissions benefits and unit costs depend greatly on the size, nature, and the context of the investment made, and generalizations are not appropriate. Some research has found that transit improvements do encourage ridership and reduce GHG emissions, but may not be enough to stem the declines in ridership that have resulted from decentralized land use, relatively inexpensive fuel until recently, and other trends of recent decades. Additionally, certain kinds of improvements—such as adding new rail lines—may produce significant GHGs that must be included in emissions accounting.

Implementation Concerns: Transit improvements can be costly (especially new heavy rail service), controversial, and may not produce the anticipated ridership gains. For cash-strapped transit agencies, building and operating new service, or even increasing the frequency of existing service, may not be feasible.

Background

Moderately- or heavily-utilized transit systems (which include bus, light rail, heavy rail, commuter rail, and paratransit) can generally transport people more efficiently, with fewer GHG emissions, than cars, particularly in comparison to single-occupancy vehicle trips. However, at the national level, transit use constitutes only 1.6% of all trips (even fewer than walking) and 1.2% of all miles traveled. Transit tends to be most popular for commute trips; about 3.4% of all trips to and from work were on transit (Hu and Reuscher, 2001). The use of transit varies significantly by region and density; about 40% of all transit trips in the US were in the New York region alone (APTA, 2009).

Transit improvement projects seek to increase transit use by increasing transit availability, convenience, and comfort. Transit improvements can include:

- adding new services, such as introducing a rail system in a region with only buses;
- adding new corridors, either rail or bus;
- increasing service along existing corridors; or

- upgrading systems and services with newer vehicles, more convenient fare payment, mechanisms (e.g., smart cards), improved information services (e.g., to let riders know when the next bus or train will arrive).

Policy and Implementing Organizations

Transit agencies—which are operated at the local, regional, and even state level—are integral to improving transit services. State DOTs also build and operate some rail lines. While many projects are undertaken with local, county, regional, and state funds, major transit improvements, such as new rail lines and extensions, are often funded in part by the Federal Transit Administration, which allocates New Starts funds to transit agencies applying for funding for capital projects. Other FTA grant programs help fund purchases of buses and upgrades to rail systems. Major capital improvements generally involve collaboration between a transit agency and other local units of government (for example, to select new rail alignments), and states may help plan and fund major transit projects.

Effects

Target Group

Transit improvements can encourage transit use among people who usually drive and further increase its use among existing riders. While rural transit systems exist, transit improvements have larger impacts on ridership and VMT in urbanized areas.

GHG Effects

Transit use is affected by many factors, which Taylor and Fink (2003) suggested are the following:

- the share of employment in a center city (transit use declines with decentralization of jobs);
- per capita income (transit use declines with rising income);
- gas prices (transit use increases when gas prices rise, although the evidence seems to indicate that the increase must be fairly large before an effect is seen);
- parking costs and availability (transit use increases when parking is scarce or costly);
- housing density (transit use increases with higher densities);
- fares (transit use increases when fares are steady or reduced); and
- service quality (transit use rises when the quality of service improves).

These do not contribute equally to ridership: one study found that internal factors (that is, service provision and fare levels) explain about one-quarter of the variation in transit ridership levels; the other three-quarters depend on external factors (Taylor et al., 2009).

Many studies have been conducted about the impact of transit availability on travel behavior, often in conjunction with land use issues. Importantly, as with land use, the issue of self-selection makes it difficult to draw conclusions about the effects of transit improvements: that is, do people change their behavior because they move to a neighborhood served by transit, or do they select a neighborhood with transit because they wish to ride transit? Most research draws comparisons between neighborhoods with similar demographic characteristics, but this does not entirely eliminate the self-selection problem.

For these reasons, the link between making improvements to transit services and achieving reductions in GHG emissions is too tenuous to be able to make inferences about how much reduction might be achieved with specific investments. Instead, this section looks at the evidence about the intermediate step—whether and by how much ridership increases with such improvements.

One study looked at the effects of light and heavy rail expansion on the use of transit for commuting between 1970 and 2000, a period that saw a large degree of suburbanization and decentralization across the U.S. (While the focus was on the effects of rail expansion, transit figures include use of both rail and buses.) Across the seven regions that had pre-existing rail lines in 1970, the average transit mode share for commuting fell from 30 to 23%. (The actual decline in mode share varied by city, but all seven cities showed decreases.) In 14 regions that added rail lines by 2000 where none existed in 1970, transit use decreased in seven, remained the same in two, and increased in five. With only one exception, the commute mode share in 2000 was 7% or less. (In comparison, in cities with only bus transit, the share of commuting by transit bus also fell, from 5 to 2%.) Essentially, the study found that the addition of rail helped stem a decline in transit use, particularly among suburban commuters. While the use of existing rail and bus lines by urban commuters fell, the new rail lines increased the number of suburban commuters—but not by enough to overcome decreases among other passengers (Baum-Snow and Kahn, 2005).

Interestingly, the study also calculates the transit use that would have been expected if decentralization had not occurred between 1970 and 2000. In all but two regions, the percentage of transit use is higher in that hypothetical, centralized case, supporting the idea that transit investments are more effective in centralized than decentralized regions (Baum-Snow and Kahn, 2005).

Another study looked at the 226 transit agencies that saw increased ridership in the second half of the 1990s. Of these, 188 agencies increased their service levels, while only 38 decreased them (service levels were measured in revenue vehicle hours,[61] and included all transit modes). The study found that ridership increased with service increases, but at declining rates of return. That is, agencies that increased their service hours the least, an average of a 4.3% increase, saw ridership

---

[61] Revenue vehicle hours are the number of hours that transit vehicles are serving passengers.

gains (in unlinked trips)[62] of 8.5%, while those that increased service the most, an average of a 79% increase, saw a 64.1% increase in ridership—large but proportionally less. While service provision contributed more to ridership gains than did fare decreases,[63] the study noted, "because the level of transit service provided is, to a large degree, a function of the demand for transit service, there is no guarantee that simply increasing service will result in corresponding ridership growth" (Taylor et al., 2002, p. 46).

One report estimated that the total fuel savings due to transit availability across the U.S. is approximately 5.2 billion gallons per year. This includes primary effects, meaning the use of transit as a substitute for private car travel, as well as secondary impacts, meaning that the more dense neighborhoods made possible by transit that are widely shown to reduce per capita VMT. This translates to CO_2 emissions "savings" of about 46 million $MTCO_2$ attributable to transit (Bailey et al., 2008).

The recent Moving Cooler report estimated that transit capital investments across the U.S. could reduce CO_2 emissions by 144 to 575 million $MTCO_2$ cumulatively by 2050. The range reflects three scenarios at different levels of aggressiveness, assuming that investments are made to increase transit ridership on all modes by 3%, 3.5%, and 4.67% starting in 2010 (Cambridge Systematics, 2009).[64] Other types of transit improvements, defined as increasing the number of revenue service miles and increasing travel speeds, would, if implemented alone, achieve less than one-third of the GHG reductions that the capital improvements would. When combined with projected decreases in emissions from investments in intercity and high-speed passenger rail, the range of possible reductions is from 0.4 to 1.1% of total GHGs from on-road transportation in the US.

BART, a heavy rail system in the San Francisco Bay Area, estimated GHG reductions for a variety of programs that it could implement with regard to transit improvements. Increasing off-peak train frequency was estimated to reduce 1,000 $MTCO_2$ per year. For rail extensions (the analysis was for a single 5.4-mile planned extension), the annual reduction could range from 30,000 to 79,000 $MTCO_2$ (the high end of this range includes land use impacts from more compact development patterns). This does not include emissions from the project's construction, which could be significant. Another 10-mile extension, using a different technology, would reduce GHG by 38,000 to 111,000 $MTCO_2$, depending on land use impacts (Nelson\Nygaard Consulting Associates, 2008).

---

[62] Unlinked trips are one-way trips on a single transit vehicle. If a rider transfers from one vehicle to another to reach a destination, each segment is one unlinked trip.

[63] Among all the agencies with gains in ridership, 68 increased fares over the study period, 67 decreased fares, and 45 held fares more or less constant.

[64] For each strategy, the report provides brief descriptions of three levels of aggressiveness: expanded best practice, more aggressive, and maximum effort. For transit capital expansion, the report stated that it would be investment sufficient to increase ridership by 3, 3.53, and 4.67% annually across modes, without reference to particular investments.

Estimated Cost per Metric Ton of CO_2 Reduction

Costs and ridership changes vary so widely from region to region that deriving a consolidated estimate of unit costs is not appropriate. Moreover, the only analysis of this relationship that appears to exist is at the national level in the Moving Cooler report. For the three capital investment scenarios, cumulative costs for capital expansion are assumed to be $256 billion, $505 billion, and $1,203 billion (2009 USD) over the period from 2010 to 2050.[65] This would imply costs of $1,770 to $2,080 per metric ton reduced. For the scenarios about other transit improvements (increasing the number of revenue service miles and increasing travel speeds), costs per ton are $1,166 to $1,451 (Cambridge Systematics, 2009).

BART's estimates included costs per metric ton reduced. For increasing off-peak service, the cost was $2,000 per metric ton. The 5.4-mile extension would cost $2,000 per ton, or $720 per ton if emission reductions were realized from land use changes. The 10-mile extension would cost $940 per ton, or $280 per ton with land use impacts. Although the report does not state it directly, the difference in cost between the two extension projects seems to be a function of the different proposed rail types.

Key Assumptions and Uncertainties

As the studies suggest, a key uncertainty is how much transit ridership will result from improvements. Taylor et al. (2002) found that in some regions, transit agencies that increased their service still saw decreases in ridership, and in the cities analyzed by Baum-Snow and Kahn (2005), the proportion of transit commuting declined in most cities with new rail investments. Clearly, increasing transit service in and of itself does not guarantee increased use. Therefore, new transit investments or services should be analyzed with respect to the specific characteristics of a region, particularly those factors that seem to affect transit use the most: the degree of centralization and land use factors such as parking and density. In addition, as with transit incentives, a number of external factors, such as gas prices, also influence mode choice.

In addition, many forms of transit improvements will produce significant GHG emissions—for example, creating new vehicles and rail lines and increasing service levels. These emissions must be considered in the life-cycle emissions of transit improvements in order to assess their effectiveness as GHG mitigation strategies. The studies cited in this report do not include these life-cycle emissions, and no estimates of the emissions associated with creating new vehicles or rail lines appear to be available.

Data and Tools

None identified.

---

[65] Figures reported in original as $255, $503, and $1,197.3 billion in 2008 USD.

U.S. Department of Transportation
Federal Highway Administration

Implementation Concerns

Agency Cost

Because costs vary so widely from project to project depending on the technology, costs of land, and assumptions about the useful life of investments, it is not possible to provide a meaningful range of costs. Rail is always more expensive than buses: costs per mile of rail construction in the study of 16 cities ranged from $7.7 to $70 million per mile for light rail and $17.7 to $365 million per mile for heavy rail (Baum-Snow et al., 2005).[66] The years of useful life for rail guideway (the right-of-way on which it operates) range from 20 years (for at-grade guideway in mixed traffic) to 125 years (for underground tunnel work). The track itself lasts from 20 to 35 years (FTA, 2010).

Increases in service by adding transit vehicles are much less costly; a new bus costs on average about $425,000, while rail vehicles cost between $1 and $3 million (APTA, 2009). For new projects, FTA assumes that rail cars have a lifespan of 25 years, and buses from 12 to 18 years (FTA, 2010). Average operating costs as measured per vehicle revenue hour[67] are $115 for buses, $189 for heavy rail, $434 for commuter rail, and $219 for light rail (FTA, 2008).

These costs must be weighed against the relative capacity of rail and bus to move people, operations and maintenance costs, and other factors.

Agency Implementation Concerns

In addition to large costs, many factors can make service improvements difficult: the need to focus on current operations rather than new projects, the difficulty of working with other units of government or interest groups, the difficulty of trying to coordinate station or corridor development with transit planning (transit agencies in the U.S. generally have little ability to influence how land around their stations will be used), and the long time frames for certain types of investments.

Social Concerns

Adding new transit service can be controversial, and attitudes depend on the location, cost, proposed fares, and proposed land uses. Arguments used by opponents against transit improvements include high costs of improvements relative to ridership or other perceived benefits, fears that better transit will bring "undesirables" to a neighborhood and reduce property values, concerns about additional traffic and limited parking in the station area, and general anti-urban sentiment. Additionally, in several particularly contentious cases, racial prejudice influenced opinions toward transit improvements, with whites opposing transit out of concerns that non-whites would more easily access their neighborhoods. A major public opinion poll about overall

---

[66] Originally reported as $7 to $63 million per mile for light rail and $16 to $330 million per mile for heavy rail, in 2005 USD.

[67] This is a standard measure in the transit industry of the hours a vehicle is in service, and accounts for fuel, maintenance, labor, and other operating costs.

attitudes toward transit found about one-third in favor, one-third opposed, and the remainder neutral (Weitz, 2008).

Other Costs/Barriers
None identified.

Interactions with other Strategies
- A strategy to encourage more compact land use may be negated if transit lines or extensions are implemented in a way that encourages low-density suburban development. In turn, strategies that are more effective in compact regions—e.g., car sharing—may also be less effective.

Unique Co-benefits
None identified.

Unique Negative Effects
- Investments in rail transit have on occasion taken resources away from buses, which can have negative equity impacts (Motavalli, 2001).

- Transit investments in areas unlikely to increase transit usage can result in unnecessary costs, encourage low-density development, and may even result in a net increase in GHG.

- Investments in new transit services, or even increased service provision, can be expensive for transit agencies, which may face budgetary shortfalls.

Where in Use
Most cities have some type of transit service, although the modes and service provision vary widely. The U.S. currently has 2,400 transit agencies operating over 5 billion vehicle miles annually (APTA 2009).

Recommendations for Further Research
The life-cycle effects of transit improvements are important but cannot be assessed without an understanding of the GHGs that are produced from transit improvements. Research should be undertaken to assess these emissions from different kinds of transit improvements.

References
American Public Transportation Association. (2009). *2009 Public Transportation Fact Book, 60th Edition.*

American Public Transportation Association. (2009). *Public Transportation Vehicle Database.*

Bailey, L., Mokhtarian, P. L., and Little, A. (2008). *The broader connection between public transportation, energy conservation and greenhouse gas reduction.* Washington, DC: American Public Transportation Association.

Baum-Snow, N., and Kahn, M. E. (2005). Effects of urban rail transit expansions: Evidence from sixteen cities, 1970-2000 [with comment]. In *Brookings-Wharton papers on urban affairs* (pp. 147-206). Washington, DC: The Brookings Institution.

Cambridge Systematics. (2009). *Moving cooler: An analysis of transportation strategies for reducing greenhouse gas emissions.* Washington, DC: Urban Land Institute.

Federal Transit Administration. (2008). 2008 National Transit Profile.

Federal Transit Administration. (2010, June 1). Standard Cost Categories workbook, revision 13. Available at http://www.fta.dot.gov/planning/newstarts/planning_environment_2580.html.

Hu, P. S., and Reuscher, T. R. (2004). *Summary of travel trends: 2001 national household travel survey* Washington, DC: U. S. Department of Transportation, Federal Highway Administration.

Motavalli, Jim. (2001). Second-Class Transit? Los Angeles' Bus Riders Union Fights for Equal Treatment. *E – The Environmental Magazine*, September/October, XII(5), online at http://www.emagazine.com/view/?1072andsrc=.

Nelson\Nygaard Consulting Associates. (2008). BART Actions to Reduce Greenhouse Gas Emissions: A Cost-Effectiveness Analysis: San Francisco Bay Area Rapid Transit District.

Taylor, B. D., and Fink, C. N. Y. (2003). *The factors influencing transit ridership: A review and analysis of the ridership literature.* UCLA Department of Urban Planning Working Paper.

Taylor, B. D., Haas, P., Boyd, B., Hess, D. B., Iseki, H., and Yoh, A. (2002). Increasing transit ridership: Lessons from the most successful transit systems in the 1990s: Norman Y. Mineta International Institute for Surface Transportation Policy Studies, MTI Report 01-22.

Taylor, B. D., Miller, D., Iseki, H., and Fink, C. N. Y. (2009). Nature and/or nurture? Analyzing the determinants of transit ridership across us urbanized areas. *Transportation Research Part A: Policy and Practice*, 43(1), 60-77.

Transport for London (TfL). (2004). *Impacts monitoring – Second Annual Report.*

Weitz, R. (2008). Who's afraid of the big bad bus? Nimbyism and popular images of public transit *Journal of Urbanism: International Research on Placemaking and Urban Sustainability*, 1(2), 157-172.

Telework

Policy: Commuting accounts for about one-third of all miles driven in the U.S. As information technology continues to improve, telework—working from home or an off-site location—has become increasingly feasible and attractive. Governments at all levels may encourage or provide incentives for employers to offer their employees the option of teleworking, thereby reducing commuter VMT.

Emissions Benefits and Costs: GHG reductions have been estimated for employees who choose to telework. Based on one national model, each teleworker reduces emissions by about 0.5 $MTCO_2$ per year. However, it is unclear how telework encouragement programs affect decisions to telework. The main cost to public agencies for telework promotion is staff time, suggesting costs in the range of a few hundred thousand dollars annually. One study suggests costs may be as little as $3-4 per $MTCO_2$.

Implementation Concerns: Such programs are generally acceptable to the public but may be resisted by employers due to concerns about management and productivity.

Background

Telework—a term generally interchangeable with telecommuting—means working from home or an alternative location closer to home. Almost one-third of the vehicle miles driven in the U.S. are to and from work, making commuting the single largest element of total vehicle travel (Hu and Reuscher, 2004). Moreover, congestion tends to peak during the hours when drivers are likely to be going to work. Therefore, some reductions in both congestion and emissions could be achieved if some employees did not drive to work and instead worked remotely (i.e., telework).

Policy and Implementing Organizations

The public sector role in telework is to encourage employers to adopt policies to allow employees to telework. Efforts to encourage greater use of telework have been undertaken at all levels of government. Localities and MPOs tend toward policies that provide outreach and technical assistance to employers, or in a few cases directly provide telework centers for employees to use. State and federal policies tend toward providing tax incentives to employers; a federal tax credit has been introduced in Congress several times but never adopted. Public sector employers have in many cases adopted telework policies for their own employees, and the Telework Enhancement Act of 2010 provides a framework for encouraging flexible work hours and telework opportunities for Federal workers.[68]

---

[68] "H.R. 1722—111th Congress: Telework Enhancement Act of 2010." GovTrack.us (database of federal legislation). 2009. September 29, 2011 http://www.govtrack.us/congress/bill.xpd?bill=h111-1722.

Effects

Target Group

Telework policies, like many TDM policies, are aimed at both employers and employees. Employers generally view telework as an employee benefit, rather than as a transportation program, and often tie it to issues unrelated to commuting, such as job description or length of service. Telework programs also target the employees who are eligible to telework (generally employees whose work can be performed away from conventional worksites).

GHG Effects

No organization regularly collects information on telework, so it is difficult to assess national trends. Most empirical studies of telework are based on small-scale programs, and only one study was found that models the nationwide impact of telework. However, there is almost no data on the effectiveness of public sector programs to *promote* telework. Where telework has been implemented and studied in the U.S., researchers have found that it leads to fewer and shorter trips among teleworkers, but that overall results have been modest, at best. The main reason is that the proportion of employees who telework is low, and these employees telework occasionally rather than full-time.

The transportation goal of telework is to reduce the number of vehicle trips or the trip length. One review of multiple studies found that on average, the VMT on a telework day decreased anywhere from 53 to 77%, and that non-commute trips did not increase (Walls and Safirova, 2004). However, since most teleworkers do not switch from driving to telework on a full-time basis, overall impacts must be gauged based on the VMT reduction of an individual driver over a longer period of time. One study of a pilot program in Los Angeles found that after two years, teleworkers worked from home on average eight days per month (Nilles, 1993). Another California study of telework centers found that the average VMT declined among teleworkers on teleworking days by 65% (38 miles). When averaged over all days, teleworking and non-teleworking, total VMT declined by 17% (Belapur et al., 1998). Both of these studies looked at the behavior of teleworkers, not at how many employees began teleworking.

A Washington, D.C. study of a short-term intensive telework promotion program designed to encourage the adoption of telework throughout the region resulted in an average reduction in VMT of 7.6 miles per day per teleworker (Ramfos and Albiero, 2006). Assuming an average car and light truck fuel economy of 20.7 MPG, this reduces about 7 lbs of CO_2 per day per teleworker.[69] The D.C. study did not provide the percentage of target employees who began teleworking, but the 4,200 new teleworkers that resulted from the program fell far short of the 113,000 goal (Ramfos and Albiero, 2006).

---

[69] The average fuel economy for the overall fleet of cars and light trucks in the US in 2006 is 20.5 mpg (calculated from Bureau of Transportation Statistics, 2009).

Perhaps the best assessment of ongoing telework assistance is at the Metropolitan Washington Council of Governments, which releases an assessment every three years of its many outreach efforts, including telework. Overall, the existing telework program in the region was estimated to account for a reduction of about 47,100 $MTCO_2$ annually over the period FY 2006-2008 (LDA Consulting et al., 2009).

One widely cited study provides an opportunity for estimating overall effects at a national level. Choo et al. (2005) used data about past telework trends and developed a model to estimate the national reduction in VMT for 1998 (that is, given what we know about how other factors affect VMT, how much less VMT is driven because of telework?). They assumed a 27-mile round-trip commute, with 76% of those miles driven alone; telework frequency of 1.2 days per week; and 15.7 million teleworkers (12% of the national workforce). The model found that telework causes total US annual passenger vehicle VMT to be 19.3 billion miles less than it would be without telework. This represents a VMT savings of 0.8%. Again, assuming an average fleet fuel economy of 20.7 mpg, this reduces GHGs by 8.2 million $MTCO_2$ annually, or about .5 $MTCO_2$ per teleworker.

Estimated Cost per Metric Ton of CO_2 Reduction

There is little literature about the effectiveness of public sector programs to promote telework, and there appears to be no literature that links the costs of public investment in telework programs to resulting declines in VMT. Most programs are implemented at the regional level through a commuter assistance program, and these programs typically do not do a detailed analysis of their outcomes. Nevertheless, the Metropolitan Washington Council of Governments telework program in the region offers one data point. It was estimated to account for a reduction of about 47,100 $MTCO_2$ annually over the period FY 2006-2008 (LDA Consulting et al., 2009). The annual cost for the region's telework program is therefore about $166,000[70] (National Capital Region Transportation Planning Board, 2006). This corresponds roughly to a reduction of about $3.50 per $MTCO_2$. The main cost component of these programs is generally staff time.

Key Assumptions and Uncertainties

Key assumptions include the number of U.S. workers whose positions are amenable to telework, the proportion who actually will take up teleworking, the average commute VMT driven by teleworkers (i.e., whether it is the same or higher than the overall commuting population), and the number of days per week teleworked. In addition, if unemployment remains high (at the time of this writing it is 8.3%) this may have long-term impacts on telework in two ways. First, the number of employees who commute may decrease as unemployment increases, and, second, if managers perceive that workers who telework are expendable or less productive, this may dampen the acceptance of telework. Choo et al. (2005) also posit that telework may have a natural plateau, or a point at which

---

[70] This figure is from the FY 2007 work plan. This has been converted to 2009 USD.

new teleworkers are balanced against those who return to commuting, for whatever reason, including changes in jobs or preferences.

Data and Tools

No single reliable source exists for determining the number of teleworkers. Various data are available from a number of sources, such as the Census, American Housing Survey, Current Population Survey, the National Household Travel Survey, and several private market research firms, and their figures do not always agree. Long-term study of telework is hampered by this lack of consistency (Mokhtarian et al., 2005). In addition, most commuter-assistance organizations do not track the impacts of their efforts to increase telework.

Implementation Concerns

Agency Cost

Operating costs for outreach programs are generally on the order of hundreds of thousands of dollars, since they largely comprise staff time. Operating costs would increase if a public sector program were to provide incentives to defray the costs incurred by employers, such as purchasing computer equipment for employees. The Telework!VA program, operated by the Virginia Department of Rail and Public Transportation, offers to reimburse qualified employers up to $35,000, provided their resulting telecommuting program meets certain benchmarks (VDRPT, 2009).

Telework centers are more expensive, but less common. Setting up telecommuting centers represents a moderate capital investment. One report recommended at least three years of public funding for a center to establish itself. Estimated total cost for a 12,000 square foot facility with 60 workstations is $1.4 million in 2009 USD($625,000 for start-up costs, plus operating costs of $22,600 per month over three years) (Bacharach et al., 2005).[71]

Agency Implementation Concerns

While there are no specific concerns associated with encouraging telework, it must be noted that agencies' abilities to influence telework habits may be limited.

Social Concerns

There are few barriers at the public or individual level to telework implementation. It is more common for employers to resist telework programs out of concerns about the difficulty of managing employees remotely. For those individuals who telework, it may be that the face-to-face interaction is too important to forgo on a daily basis, which helps explain why it is more common to telework on occasion rather than daily (Rosenberg, 2007).

---

[71] This study originally reported $500,000 in start-up costs and $18,000 per month in operating costs in 2000 USD. This has been converted to 2009 USD.

Other Costs/Barriers

One study estimated the employer cost to establish a telecommuting program for their employees is roughly $3,000 in one-time costs and $1,100 in recurring costs. These costs include computer equipment and the associated telecommunications upkeep (JALA, 2009). Some evidence suggests that employers with teleworking employees experience some productivity gains, which may offset these costs (Butler et al., 2007), but this continues to be a topic of debate in the literature (Bailey and Kurland, 2002).

Interactions with other Strategies

- While telework can be encouraged on its own, it is one of many employer-based TDM strategies. These strategies may have a larger collective effect and be less expensive to publicize when promoted together rather than individually.

Unique Co-benefits

- Teleworkers may experience lower driving costs.

Unique Negative Effects

None identified.

Where in Use

Telework is fairly widespread in both the public and private sectors, although, as noted above, no comprehensive estimates or databases exist.

Recommendations for Further Research

It is possible that other regions in addition to Washington, DC might have assessed the cost and effectiveness of telework promotion programs, and a broader review of project reports would be valuable.

References

Bacharach, Jacki and Siembab, Walter. (2005, June 30). *Implementation steps for two strategies promoting jobs/housing balance: Local preference housing and share work/communications centers.* Final Report. Prepared for Ventura Council of Governments.

Bailey, D. E., and Kurland, N. B. (2002). A review of telework research: Findings, new directions, and lessons for the study of modern work *Journal of Organizational Behavior*, 23(4) Special Issue: Brave New Workplace: Organizational Behavior in the Electronic Age, 383-400.

Belapur, Prashent, Varma, Krishna, and Mokhtarian, Patricia. (1998). "The transportation impacts of center-based telecommuting: interim findings from the neighborhood telecenters project." *Transportation*, vol. 25, no. 3.

Bureau of Transportation Statistics (2009). *National Transportation Statistics, 2009.* U.S. Department of Transportation. Available online at: http://www.bts.gov/publications/national_transportation_statistics/.

Butler, E. S., Aasheim, C., and Williams, S. (2007). Does telecommuting improve productivity? Seeking solid evidence of demonstrable productivity gains. *Communications of the ACM*, 50(4), 101-105.

Choo, S., Mokhtarian, P. L., and Salomon, I. (2004). Does telecommuting reduce vehicle-miles traveled? An aggregate time series analysis for the U.S. *Transportation*, 32(1), 37-64.

JALA International. (2009). Home-based Telecommuting Cost-Benefit Analysis, web page available at www.jala.com/homecba.php, accessed December 15, 2009.

LDA Consulting, CIC Research Inc., ESTC, and Center for Urban Transportation Research. (2009).

Transportation emission reduction measure (TERM) analysis report, FY 2006-2008. Washington, DC: National Capital Region Transportation Planning Board Commuter Connections Program.

Mokhtarian, P., I. Salomon, and S. Choo. (2005). "Measuring the measurable: Why can't we agree on the number of telecommuters in the U.S." Quality and Quantity 39(4): 423-452.

National Capital Region Transportation Planning Board Metropolitan Washington Council Of Governments. (2006). Work Program For The Commuter Connections Program For The Greater Washington Metropolitan Region, Fiscal Year 2007.

Nilles, Jack (1993, March*). City of Los Angeles telecommuting project: Final report.*

Ramfos, Nicholas and Albiero, Rick. (2006). *Expanded telecommuting transportation emissions.*

reduction measure (TERM) final report. Prepared for Metropolitan Washington Council of Governments. January 7, 2006. Available online at www.mwcog.org/store/item.asp?PUBLICATION_ID=258.

Rosenberg, M. (2007). Slow but steady "Telework revolution" eyed. *Cascadia Prospectus.*

Virginia Department of Rail and Public Transportation. (2009). Telework! VA web page, available at www.teleworkva.org/, accessed December 15, 2009.

Walls, M. and E. Safirova (2004). A Review of the Literature on Telecommuting and Its Implications for Vehicle Travel and Emissions. Washington, D.C., Resources for the Future.

6. Transportation System Management Strategies

Transportation system management (TSM) refers to a set of strategies that largely aim to reduce GHG emissions by reducing congestion, primarily by improving transportation system capacity and efficiency. TSM strategies may also address a wide range of other externalities associated with driving such as pedestrian/driver safety, efficiency, congestion, travel time, and driver satisfaction. Some TSM strategies are designed to reduce total and systemic congestion and improve system-wide efficiency, while other strategies target particularly problematic areas where improvements could greatly affect congestion, safety, efficiency, and GHG emissions.

Transportation System Management Strategies Reviewed in This Report

This review covers the following eight reduction strategies:

Traffic Signal Optimization 107

Ramp Metering 117

Incident Management 125

Speed Limit Reduction and Enforcement 133

Roundabouts 141

Capacity Expansion 149

Resurfacing Roads 155

Alternative Construction Materials 163

The above strategies seek to reduce congestion and promote efficiency through infrastructure, operational, and technological improvements. Infrastructure strategies seek to reduce GHG emissions by improving the transportation system infrastructure through new or improved construction. The production of pavement materials requires significant amounts of energy and produces significant GHGs. "Green" construction materials are lower-energy alternatives to conventional construction materials and can reduce the life-cycle GHG emissions of transportation construction projects. Resurfacing roads decreases the roughness of road surfaces and allows vehicles to travel more efficiently, which in turn reduces GHGs. Capacity expansion and roundabouts may reduce congestion and allow for more free-flowing traffic with less stoppage and idling time.

Operational strategies focus on minimizing inefficient travel that increases GHG emissions. Proper incident management strategies detect and clear incidents to reduce congestion and promote safer post-accident operations. Speed enforcement and reduction programs seek stricter enforcement of speed regulations as well as lower limits so that travel speeds coincide with speed ranges that promote optimal fuel efficiency.

Technological strategies seek to use automated systems to optimize free-flow of traffic and thus reduce non-VMT production of GHGs. Traffic signal optimization improves the operation, maintenance, timing and location of traffic signals to promote smoother traffic flow and reduce GHG emissions. Ramp metering controls the rate of vehicles entering freeways to reduce congestion around ramps and discourage use of highways/freeways for short trips.

Current FHWA research efforts address the short- and long-term impact of highway operations on travel and GHG emissions. Strategies of particular interest include signal timing, ramp metering, incident management, speed harmonization, and congestion pricing. The travel behavior component of this work will examine key factors affecting travelers' responses to these treatments, such as demand changes from changes in travel time, travel time variability, and travel cost. An important outcome of this research will be to characterize what we know about travel in the months and years following implementation of these strategies. Ultimately, this work will support travel experiments to estimate the network-level travel and GHG impacts of individual and bundled highway measures.

Conclusions Regarding TSM Strategies

States, regions, counties and municipalities have implemented many TSM strategies over the years because they address a wide range of externalities associated with transportation operations. Recently, TSM strategies have been proposed as a way to address GHG emissions in particular.

However, the literature on the GHG effects of the TSM strategies reviewed in this study is largely inconclusive. There are several reasons for this. First, most TSM strategies seek to improve the efficiency or capacity of the transportation system, which enables people to travel at higher speeds and with less congestion, reducing the time and comfort costs of driving. Reducing the cost of driving also induces demand, which may not be characterized in studies of GHG effects. For example, some research on capacity expansion has shown that it significantly induces demand and may result in a net increase in GHG emissions. This may be particularly so in growing areas where capacity increases attract further development or change development patterns. This suggests that TSM strategies should be closely coordinated with urban planning and other land use concerns. Other TSM strategies that one would expect to be affected by induced demand are road resurfacing, traffic signal optimization, and incident management, though it may be less significant than for capacity expansion.

Second, many TSM strategies seek to improve system performance through construction projects. This includes road resurfacing, roundabout construction, and capacity expansion. The construction process itself emits significant GHGs, and the life-cycle emissions are largely not taken into account in the literature. For example, research suggests that road resurfacing emits 9.7 tons of CO_2 per lane mile for the construction alone (see discussion of road resurfacing) which can reduce or negate the fuel efficiency benefits of a smoother road, particularly when coupled with the higher speeds that smooth roads enable. Constructing new capacity can produce several *thousand* tons of CO_2 per lane

mile. Without knowing the life-cycle emissions, one cannot know the GHG reductions that are possible.

Induced demand and unaccounted-for life-cycle emissions are both examples of unintended GHG emissions that arise from these strategies. There are other kinds of unintended GHG emissions as well. As noted, some of these strategies enable faster driving. At speeds higher than approximately 55mph, however, GHG emissions from vehicles increase significantly. This has been shown to offset some of the gains from improving flow, for example with ramp meters and road resurfacing. Additionally, while ramp metering improves flow on highways, it increases idling at the ramp, which also produces GHGs. These unintended emissions also reduce the effectiveness of these strategies.

Because of these factors, the evidence for road resurfacing, capacity expansion, roundabouts, and ramp metering is mixed – some studies show a reduction in emissions while others show an increase in emissions. This is not to suggest that they should never be used; in some cases, they may indeed be effective at reducing GHG emissions. For example, targeted capacity expansion at existing bottlenecks could reduce congestion, with limited induced demand. Rather, the evidence speaks to a need for carefully assessing *all* of the sources of GHG emission reductions and increases, both intended and unintended.

Two strategies may be particularly promising in reducing GHG emissions and counteract sources of unintended emissions. The reduction and stricter enforcement of highway speed limits could significantly reduce GHG emissions because they encourage drivers to travel at more fuel-efficient speeds, typically between 45 and 55 mph. Their effect on congestion will vary depending on the context. Where congestion is increased or travel is delayed, GHG emissions may increase because of stop-and-go traffic but may also decrease because of reduced demand. Where congestion decreases, the inverse may be true. Research suggests that speed reductions overall are likely to reduce GHG emissions. Speed reduction is likely to interact positively with other TSM strategies that would otherwise encourage faster speeds. This strategy would, however, require active support from state DOTs, DMVs, law enforcement officials, communities, and drivers, and may be unpopular among drivers who are accustomed to driving at high speeds.

The use of "green" construction materials could also significantly reduce GHGs relative to other strategies because it offsets the use of GHG-intensive materials that would otherwise be used, without affecting capacity, efficiency, or speed that would induce demand. Using green construction materials in other construction-based GHG mitigation strategies, such as road resurfacing or capacity expansion, would help to reduce the life-cycle emissions from these strategies and may be necessary in some cases to achieve net reductions.

Like transportation demand management strategies, TSM strategies have the most significant effect on GHGs when the emissions from driving are high. For example, as vehicle engines improve to be more efficient at a wider range of speeds, the absolute GHG effect of speed reduction programs will decline. This implies that the effectiveness of TSM strategies may decline as vehicles and fuels improve. All of these strategies are important in combating climate change, but their combined effect will be less than the sum of their individual effects. Additionally, TSM strategies may undermine the effectiveness of TDM strategies by reducing the cost of driving and therefore inducing demand. The exception is fuel taxes and road pricing which would counteract the induced demand from improved system performance because they make driving more expensive and reduce VMT.

Finally, TSM strategies have common co-benefits and negative effects. Strategies that encourage faster driving such as traffic signal optimization may reduce safety for pedestrians, cyclists, and also drivers. Those that calm traffic, such as speed reduction programs and roundabouts, may improve safety. Strategies that decrease travel times and congestion, or improve driving comfort, are likely to improve driver satisfaction. Those that improve the efficiency of the system may increase travel time reliability or satisfy greater demands without the need for added capacity, which is both costly and may induce demand still further.

Traffic Signal Optimization

Policy: Traffic signals can increase stop-and-go driving, causing aggressive acceleration and deceleration, congestion, and excess idling, all of which reduce fuel efficiency and increase GHG emissions. Traffic signal optimization is the process of improving the operations, maintenance, timing, and location of traffic signals to promote smoother traffic flow, which simultaneously reduces GHG emissions.

Emissions Benefits and Costs: Where traffic signal optimization has been implemented and studied, the literature shows 3-12% fuel savings and GHG emissions at signalized intersections. These results must be interpreted cautiously, however, because, like other strategies that improve traffic flow, signal optimization may induce demand and reduce the stated benefits, something not clearly accounted for in the studies cited. The literature also suggests that when optimization is undertaken at intersections that are already signalized, costs may range from $25 to $34 per $MTCO_2$.[72]

Implementation Concerns: Signal optimization is often undertaken to improve traffic flow, and reductions in GHG emissions are seen as an added benefit. Given this, signal optimization is likely to be supported by the public. Coordination across jurisdictions may be necessary, but challenging, for large signal optimization projects.

Background

Traffic signals can increase stop-and-go driving, causing sudden acceleration and deceleration, congestion, and excess idling, all of which reduce fuel efficiency and increase GHG emissions. Traffic signal optimization is the process of improving the operations, maintenance, timing, and locations of traffic signals to promote smoother traffic flow and mitigate these effects.

A key traffic signal optimization tactic is the coordination of signals (i.e., the length of green and red signals and the timing of signal changes) in a corridor to maximize green light time for vehicles traveling at the speed limit. This creates smoother traffic along the corridor. Such optimization can be static or dynamic. In static optimization, signals are timed to operate according to a fixed schedule, while dynamic optimization uses real-time traffic data to adapt signal timing.

Other traffic signal optimization tactics include:

- removing unneeded signals;
- adding traffic detectors to side streets, thereby enabling traffic progression through the system;
- installing new signal equipment, such as solid state electronic controllers, that provide the capability to implement more advanced (e.g., dynamic) traffic control; and

---

[72] All values updated to be consistent with 2009 USD, unless noted.

- controlling signals from a central location, thereby enabling remote management.

The literature, and hence this review, focuses almost entirely on signal coordination approaches to traffic signal optimization.

Policy and Implementing Organizations

Traffic signal optimization is typically undertaken at the local and regional level and can involve the retiming of one intersection or the coordination of signals at multiple intersections. MPOs typically coordinate signal optimization projects, while DOTs primarily provide funding, and local jurisdictions implement and maintain the lights. Many local agencies may need to coordinate when signal optimization projects span multiple jurisdictions.

Effects

Target Group

This strategy affects roads with signalized intersections and indirectly affects the behavior of drivers.

GHG Effects

Traffic signal optimization benefits (including emissions and fuel reductions) have been studied since the early 1980s. Earlier system models primarily estimated fuel savings from decreases in congestion. More recently, models have evolved to take into account the effects of reduced acceleration and deceleration to provide a more complete estimate of the effects on fuel consumption and, therefore, GHG emissions. Although GHG reduction estimates appear to be consistent, most of the practical research has been conducted in large cities and the results cannot necessarily be generalized to other areas. In addition, the benefits of traffic signal optimization vary depending on a number of factors, including the previous level of congestion and traffic, and the specifics of the optimization approach. Finally, the improvement in traffic flow may induce demand, which is not explicitly taken into account in most studies, but which may reduce the benefits of signalization.

A few studies report the fuel consumption effects per signalized intersection:[73]

- A study in Toronto, Canada found that traffic signal optimization reduces fuel consumption by 4-7% and reduces emissions of all vehicle pollutants (including, but not limited to, CO_2) by 3-6% per intersection within the study area. Based on the positive results of the program, Toronto expanded the system to control approximately 250 additional signalized intersections. When researchers accounted for reduced personnel costs associated with ongoing system operation (e.g., it was estimated that six additional staff members would be required to develop the large number of additional fixed-timing plans to partially replicate

---

[73] Many studies do not explicitly state the spatial area that constitutes an intersection or the extent of the study areas and corridors.

the level of on-street performance that the automated system can provide within the 75-signal demonstration), they calculated that the payback period to the City of Toronto for the monetary cost of the expansion project was under two years (Greenough and Kelman, 1999).

- A Parisian study (Midenet et al., 2004) found that adaptive real-time signal control system CRONOS led to a 3-4% reduction in GHG emissions at the site of each intersection. An on-field experiment was performed at an isolated signal intersection during 8 months in a close-in suburb of Paris to evaluate the benefits of the CRONOS control system. The CRONOS system (described by Boillot et al. [1992]) continuously reacts to ongoing local traffic conditions through video measurements, such as queue-lengths, and adapts the signals to minimize delay. Using video sensors to measure speed and volume of all traffic going through the intersection, Midenet et al. (2004) determined that signalization reduced an average of 8.8 lbs and 17.6 lbs of CO_2 per intersection per hour during off-peak and peak hours, respectively.[74] Peak hour traffic in the intersection was between 2,600 and 3,300 vehicles per hour.

Other studies, dating as far back as the 1980s, offer project-wide estimates of fuel or CO_2 savings:

- In 1983, 41 California cities retimed 1,535 signals. Field studies reported reduced vehicular delays and fuel consumption and the program was expanded. Over the next 11 years, 160 California cities and counties retimed 12,245 signals in 334 projects. Fuel consumption through these improved areas have been reduced by 8% (Berkeley, 1994). In the first year, reductions resulted in annual savings of approximately 6.4 million gallons of fuel (California Energy Commission, 1984). This equates to 56,898 metric tons of CO_2 (0.06 million $MTCO_2$) annually and 37 $MTCO_2$ per year per intersection. Given advancements in signal technology and traffic models in the last 25 years, this may be lower than the reductions that would be possible in the same scenario today.

- As part of the Clinton Climate Initiative (2009), the City of Portland optimized traffic signal timing at 135 intersections on 16 city streets. This optimization work has saved motorists over 1,750,000 gallons of gas each year. This reduction in gasoline consumption is equivalent to 15,460 $MTCO_2$ annually and 115 $MTCO_2$ per year per intersection.

A few regional studies reported a percentage of average fuel savings for a series of intersection improvements. While this is equivalent to the percent CO_2 savings, it does not inform us about absolute CO_2 savings:

- California's Fuel Efficient Traffic Signal Management (FETSIM) program optimized 3,172 traffic signals through 1998, and reported an average reduction in fuel use at these intersections of 8.6% for the program (Skabardonis, 2001).

---

[74] This was originally reported as 4 and 8 kg of CO_2 for off-peak and peak hours, respectively.

- A project that optimized 700 signals in the Tysons Corner area of Northern Virginia resulted in a reduction in fuel use of 10-12% for those intersections (White et al., 2000).

- A study of signal optimization of 223 signals along seven corridors in Nashville, Tennessee found a fuel consumption reduction of nearly 6% along the seven corridors (Kimley-Horn and Associates, 2006).

Estimated Cost per Metric Ton of CO_2 Reduction

Traffic signal optimization costs include hardware (traffic detectors and new signal equipment, such as solid state electronic controllers), maintenance, signal timing plans, and remote capabilities to manage signals. Recent estimates show that optimization of existing signals costs between $2,600 and $4,000 per intersection (NTOC, 2007; Kittelson and Associates, 2008; Sunakari, 2009). Considering that traffic signals should be retimed every three years (NTOC, 2007), this results in approximate costs of $1,000 to $1,300 per year per intersection. The annual CO_2 reduction estimates of 37, 39, and 115 metric tons per intersection observed in the literature suggest costs from between $8 and $35 per metric ton of CO_2. This does not consider the societal benefits from signal optimization, including fuel savings to drivers, costs of delay, and other potential savings. However, this also does not consider the costs of installing new traffic signals, which can range from $86,600 to $202,000 (Sorensen et al., 2008).

Key Assumptions and Uncertainties

There are several sources of uncertainty in analyzing traffic signal optimization. First, efficiencies in the system may induce greater demand for vehicular traffic and create a rebound effect, which would not be reflected in short-term studies of intersection performance. The CCAP guidebook (2006) assumes a 20% rebound effect, and most studies do not articulate the effects of induced demand from signal optimization (these include Boillot et al., 1992; Boillot et al., 2000; Midenet, et al., 2004; Pandian, et al., 2009; Rahka at al., 2000). It would be possible for an area considering this strategy to review historical traffic count data for "before" and "after" conditions in corridors where signal optimization has been applied in the past. However, such analysis would also need to account for exogenous factors such as changes in population and employment, fuel prices, and diverted demand from other facilities.

Second, estimates and models simplify patterns of fuel consumption and vehicle travel and cannot account for the wide array of vehicles in operation. Even basic differences between vehicles, such as make and age, are simplified in models, affecting the outcomes of the studies (Stevanovic, 2009).

Third, an important aspect of the cost/benefit analysis is whether it includes travel delay costs. Considering delay costs and fuel savings in the analysis increases the cost effectiveness of this strategy (since signal optimization reduces delay and fuel consumption). One study estimated that signal coordination would save 19.6 million hours of delay and save $418,751,000 if implemented in all 429 urban areas (Schrank and Lomax, 2009).

Data and Tools

A variety of software can aid in implementing signal optimization across several intersections. Older techniques such as TRANSYST (Roberston, 1969) optimize signals by pre-timing them. New techniques (e.g., CRONOS or the Sydney Coordinated Adaptive Traffic System (SCATS)) use real time data to "match" the current traffic conditions to the "best" pre-calculated off-line timing plan (Yu and Recker, 2006).

Similarly, a variety of commercial modeling systems can be used to determine delay, fuel consumption, and emissions from signalized intersections. These include, for example, SYNCHRO and TRANSYT-7F (for traffic flow), VISSIM-CMEN-VISCAOST (for scenario based fuel consumption), aaMOTION (a single vehicle software package for modeling fuel, emissions, and costs), PASSER II (measures cycle lengths in algorithm to estimate delay) and aaSIDRA (an intersection analysis software package).[75] If changes in vehicular travel activity can be measured or modeled, EPA's MOVES model can also be used to estimate changes in emissions.

Implementation Concerns

Agency Cost

Transportation and public works agencies (local, regional, and state levels) use local, regional, state, and federal funds to undertake traffic optimization. Major costs include hiring specialists internally or as consultants to implement traffic optimization plans, and obtaining software and signalization technology.

Certain technologies such as fiber-optic networks to relay real-time traffic information may be costly, yet overall most traffic signalization projects are not considered very expensive—$2,600 to $4,000 per intersection (NTOC, 2007; Sunakari, 2009)—assuming that modern signals already exist at the intersections in question. However, according to the Institute of Transportation Engineers, signal re-timing should be considered no less than every three years—and preferably every year—to take into account new traffic patterns and demands (Sunakari, 2009). Overall, signal optimization technology has been proven, the cost is relatively low, and agencies are familiar with the implementation methods (TTI, 2007). These costs are for optimization of the signals and not the implementation costs of installing new signals.

Agency Implementation Concerns

In areas where corridors span several jurisdictions (e.g., city or county lines), achieving effective signalization may require agreements and coordination between multiple agencies.

Social Concerns

---

[75] These products are included as examples of available tools, but they have not been vetted nor are they are endorsed specifically.

U.S. Department of Transportation
Federal Highway Administration

There is likely to be little opposition to these programs given the added benefits of reduced congestion and shorter travel times (Sorensen et al., 2008).

Other Costs/Barriers

Signalization may raise concerns about the negative impacts on pedestrian/bike crossings if longer green lights lead to shorter and fewer crossings opportunities.

Interactions with other Strategies

- Ramp meter signals have the potential to conflict with arterial traffic signals, and traffic signal optimization can help synchronize the signals to make ramp meters more effective (Sorensen et al., 2008).
- When deciding on a strategy to improve delay at intersections, agencies must decide between traffic signal optimization and roundabouts, which have similar goals but use different mechanisms that may not be complementary.

Unique Co-benefits

The most obvious co-benefit is travel time savings due to smoother traffic flow and costs savings due to lower fuel consumption.

Additional benefits include:

- improvements in congestion, which benefits commercial, emergency and public transit vehicles;
- greater reliability in travel times and reduced delay (FHWA, 1995);
- reduction in aggressive driving behavior (New York State Department of Environmental Conservation, 2010); and
- less need for additional capacity (Maricopa Association of Governments, 2010).

Unique Negative Effects

If a traffic signal optimization project reduces the pedestrian and bike crossing time to improve motor vehicle flow, pedestrians and bicyclists' usability of such crossings could decrease, potentially limiting the walkability and bicycle-friendliness of the area.

Where in Use

Use is widespread, although in most places the impetus is more likely improvements in traffic flow than emissions reduction. Some specific examples include:

- Los Angeles (Sorenson et al., 2008);
- France (Midenet, 2004; Boillot, 2000);

- Virginia (White, et al., 2000);
- Toronto, Canada (Greenough and Kelman, 1999);
- California (FHWA, 1995);
- China (Pandian, 2009; Li et al., 2004);
- Nashville, TN (Kimley-Horn and Associates. 2006);
- Florida (Stevanovic, 2009); and
- Portland, OR (http://www.c40cities.org/bestpractices/transport/portland_traffic.jsp).

Recommendations for Further Research

It would be beneficial to address the many uncertainties associated with traffic signal optimization, particularly induced demand caused by improved mobility, to account for the full effect of these projects.

References

Berkeley, University of California. Evaluation of the Fuel-Efficient Traffic Signal Management (FETSIM) Program: 1983-1993.

Boillot, F., Blosseville, J.M., Lesort, J.B., Motyka, V., Papageorgiou, M., Sellam, S. (1992). Optimal signal control of urban traffic networks. In: 6th IEE International Conference on Road Traffic Monitoring and Control, London.

Boillot, F., Midenet, S., Pierrel_ee, J.C., (2000). Real-life Cronos evaluation. In: 10th IEE International Conference on Road Traffic Information and Control, London.

California Energy Commission (1984). Fuel-Efficient Traffic Signal Management: Results from the 1983 Program.

Center for Clean Air Policy. (2006). Transportation Emissions Guidebook. http://www.ccap.org/safe/guidebook/guide_complete.html.

Climate Clinton Initiative.(2009). http://www.c40cities.org/bestpractices/transport/portland_traffic.jsp.

FHWA. (1995). Improving Traffic Signal Operations: A Primer. Prepared by the Institute of Transportation Engineers for the Federal Highway Administration, U.S. Department of Transportation.

Garber, N.J., Hoel, L.A. (2002). Traffic and Highway Engineering. The Wadsworth Group, Pacific Grove.

Greenough and Kelman. (1999). ITS Technology Meeting Municipal Needs - The Toronto Experience. Paper presented at the 6th World Congress Conference on ITS. Toronto, Canada.

Kimley-Horn and Associates. (2006). Traffic Signal Optimization Study for the Metro Nashville Signal System. Prepared for the Metropolitan Government of Nashville and Davidson county Department of Public Works.
http://www.nashville.gov/pw/pdfs/Signal_Optimization_FINAL_2006.pdf.

Kittelson and Associates. (2008). Signal Timing Manual. Prepared for the Federal Highway Administration. FHWA-HOP-08-024 http://www.signaltiming.com/.

Li, Xiugang; Li, Guoqiang; Pang, Su-Seng; Yang, Xiaoguang; and Tian, Jialin. (2004). Signal timing of intersections using integrated optimization of traffic quality, emissions and fuel consumption: a note. Transportation Research Part D 9 pp. 401–407.

Maricopa Association of Governments. (2010). Traffic Signal Optimization Program, http://www.azmag.gov/Projects/Project.asp?CMSID=1050&CMSID2=1138.

Midenet, Sophie; Boillot, Florence; Pierrelee; Jean-Claude. (2004). Signalized intersection with real-time adaptive control: On-field assessment of CO_2 and pollutant emission reduction. Transportation Research Part D 9 pp. 29–47.

National Transportation Operations Coalition. (2007). National traffic signal report Card Executive Summary.

New York State Department of Environmental Conservation. (2010). Reduce Municipal Energy Use for Transportation, Local Action Overview (Transportation), http://www.dec.ny.gov/energy/56925.html#optimizing.

Pandian,Suresh; Gokhale, Sharad; Ghoshal; and Aloke Kumar. (2009). Evaluating effects of traffic and vehicle characteristics on vehicular emissions near traffic intersections. Transportation Research Part D 14 (2009) 180–196.

Rakha, H., Aerde, M.V., Ahn, K., Trani, A. (2000). Requirements for evaluating traffic signal control impacts on energy and emission based on instantaneous speed and acceleration measurements. Transportation Research Record 1738, 56–67.

Robertson, D.I. (1969). TRANSYT: A traffic network study tool. RRL Report LR 253, Road Research Laboratory, England.

Robertson, D.I., Lucas, C.F., Baker, R.T. (1980). "Coordinating traffic signals to reduce fuel consumption." *Transport Research Laboratory (TRL) Report – LR934*, Crowthorne, Berkshire, U.K.

Rouphail et al. (2000). "Vehicle Emissions and Traffic Measures: Exploratory Analysis of Field Observations at Signalized Arterials": http://www4.ncsu.edu/~frey/emissions/trb2001paper.PDF.

Schrank, David, and Lomax, Tim. (2009). Texas Transportation Institute, Urban Mobility Report. The Texas A&M University System. http://mobility.tamu.edu.

Skabardonis, Alexander. (2001). ITS Benefits: The Case of Traffic Signal Control Systems. Paper presented at the 80th Annual Transportation Research Board Meeting. Washington, District of Columbia.
http://www.itsbenefits.its.dot.gov/its/benecost.nsf/ID/42419C3E5993E9CD852569EA0071D556.

Sorensen, et al. (2008). Moving Los Angeles: Short Term Policy Options for Improving Transportation. RAND Corporation. Santa Monica, CA.

Stevanovic, A, Stevanovic, J, Zhang, K and Batterman, S. (2009). Optimizing Traffic Control to Reduce Fuel Consumption and Vehicular Emissions: An Integrated Approach of VISSIM, CMEM, and VISGAOST. TRB 2009 Annual Meeting CD-ROM.

Sunakari, Srinivasa. (2009). Benefits of Retiming Traffic Signals. ITE Spring Conference 2009. http://www.ite.org/meetcon/2009TC/Session%2012_Srinivasa%20Sunkari.pdf.

Victoria Transportation Policy Institute TDM Encyclopedia. http://www.vtpi.org/tdm/tdm59.htm.

White, et al. (2000). Traffic Signal Optimization for Tysons Corner Network Volume I. Evaluation and Summary. Published By: Virginia DOT. Report No. TPE.R7D.03.08.00.
http://www.itsbenefits.its.dot.gov/its/benecost.nsf/ID/5526CCA9B17F9309852569610051E2DA?OpenDocument&Query=Home.

Yu, X. H. and Recker, W.W. (2006). Stochastic adaptive control model for traffic signal systems. Transportation Research Part.

Ramp Metering

Policy: Ramp meters control the rate of vehicles entering the freeway in order to create more space between vehicles so that they do not collide or disrupt the highway traffic flow. They reduce congestion on the freeway but increase idling on the ramp, both of which affect fuel consumption and CO_2 emissions.

Emissions Benefits and Costs: The benefits from ramp metering to reduce GHG emissions are uncertain. Some studies report decreases in CO_2 emissions, primarily from smoother traffic, while others report increases, in part because of idling at meters. Costs also range widely. One study estimated approximately $267,000 per ramp; another estimated that it annually costs $7,650 per ramp,[76] which includes installation, maintenance, and operational costs. The costs per unit of GHG reduction cannot be estimated.

Implementation Concerns: Ramp metering may be expensive, and the public may oppose it due to delays at the ramp and perceptions of inequity.

Background

Ramp meters control the rate of vehicles entering the freeway in order to create more space between vehicles entering the freeway so that those vehicles do not collide or disrupt the highway traffic flow. This is achieved through the use of queue detectors and traffic signals at freeway on-ramps to allow only one vehicle to enter the freeway per some short time interval (e.g., every five seconds). Ramp meters allow freeways to accommodate more vehicles with fewer collisions and greater reliability (TTI, 2007). They also reduce the number of entering vehicles by encouraging drivers to use parallel streets for short distance trips in order to avoid the ramp wait time (Cambridge Systematics, 2001). In these ways, ramp meters typically reduce congestion on the freeways.

The effects of ramp meters on fuel consumption and emissions are unclear. The reduced congestion on the freeway allows for greater fuel efficiency and reduced emissions once on the throughway. However, the decrease in congestion may increase speeds and induce demand. Vehicles idling at ramp meters and then accelerating from a full stop also have increased rates of fuel consumption and emissions. In addition, most ramp meters react to, rather than predict, bottlenecks (Pearson et al., 2003); the time delay between detection and corrective action can cause traffic fluctuations.[77] All of these factors may reduce or negate GHG benefits from reduced congestion.

---

[76] All values consistent with 2009 USD, unless noted.

[77] Researchers are now using predictive algorithms to predict traffic and potentially delay or prevent bottleneck formation (Bogenberger et al., 2001).

Policy and Implementing Organizations

Ramp meters may be installed by local/municipal, regional (MPO), and state transportation agencies.

Estimated Effects

Target Group

Ramp meters affect highway traffic, and may also affect local traffic by re-routing some trips to local roads.

GHG Effects

While there are several studies of the effects of ramp metering on congestion and safety, only a few studies consider the impacts of ramp meters on fuel consumption and CO_2 emissions. Moreover, these studies have varying results. Some practical studies of ramp metering systems have confirmed that ramp meters increase fuel use and GHG emissions, while other studies, including theoretical research, have found that ramp meters can decrease fuel consumption (Bogenberger et al., 2001; Piotrowictz and Robinson, 1995; Oregon DOT, 1982).

The literature on the effects of ramp meters varies due in part to differences in what is actually accounted for in the studies; for example, whether idling at ramps, induced demand, or increased speeds from improved traffic flow are considered.

- A study of the Twin Cities region in Minnesota found that ramp metering improved traffic volume, travel time, travel time reliability, safety, and particulate emissions on highways. However, it worsened annual fuel consumption by 5.5 million gallons of fuel and produced approximately an additional 50,000 metric tons of CO_2 (Cambridge Systematics, 2001). This increase was due to the increased speeds on highways and the time vehicles spent idling at ramp meters. The change to vehicles' speed profiles resulted in a net decrease in the emission of hydrocarbons and carbon monoxide while elevating the emission of NOx and increasing overall fuel consumption.

- The Oregon DOT installed 16 fixed-time metered ramps between downtown Portland and the Washington State line in 1981 and operated these during the morning and afternoon peak periods. Fourteen months after installation, benefits for the afternoon peak period were measured and compared to indicators before the installation of ramp meters. It was estimated that fuel consumption during the afternoon peak period, including the additional consumption caused by ramp delay, was reduced by almost 450 gallons of gasoline per weekday (Piotrowicz and Robinson, 1995; Oregon DOT, 1982). This translates to approximately 4 metric tons of CO_2 reduced daily, and 1000 metric tons reduced annually from workdays alone (assuming 250 workdays per year). The study also reported that average speeds increased from 64 mph to 69 mph. One can hypothesize that this study implicitly accounts for induced demand given that effects were measured fourteen months after ramp meter installation. It is unknown, however, whether the fuel loss from higher speeds was factored into the fuel savings analysis.

- One study simulated different ramp control algorithms for a particular 26 km (16.2 mile) stretch of freeway in Munich, Germany. It found that all control algorithms reduced fuel consumption by an average of 25%[78] (Bogenberger et al., 2001).

Collectively, this literature suggests that CO_2 effects are ultimately unknown but may vary greatly, from positive to negative net effects.

Estimated Cost per Metric Ton of CO_2 Reduction

No estimated costs per metric ton of CO_2 were reported in the literature. Moreover, this cannot be reliably calculated given that few studies report costs of ramp meters and that CO_2 effects are uncertain and vary greatly, from both positive to negative net effects.

Key Assumptions and Uncertainties

Ramp meters affect fuel consumption in ways that are often not included in calculations or models, including:

- excess fuel consumption from queues at the ramp meters due to increases in idle time and acceleration to enter the freeway;

- speed increases due to improved traffic flow;

- emissions from necessary on-ramp improvements (e.g., ramp striping) that are needed to facilitate and maximize the effectiveness of the meters (DKS Associates, 2008); and

- induced demand from reduced congestion.

As a result, the effectiveness of ramp meters as a mitigation strategy is unknown.

Data and Tools

Traditionally, ramp meters use fixed-time or traffic-responsive algorithms, which not only have the same features as fixed-time meters, but also have some ability to adapt to current conditions. Increasingly, sophisticated system-wide adaptive ramp metering (SWARM) algorithms that account for real-time traffic conditions are being used (Ahn, et al., 2007), although this requires a computerized communication center to calculate real-time adjustments along with a communication system to relay adjustments back to ramp meters (Sorensen, 2008).

---

[78] Local speed, traffic flow, and occupancy on the mainline were measured immediately upstream of the on-ramp merge.

One limitation is that most modeling techniques have not been capable of capturing off-cycle conditions (e.g., hard accelerations) and, in turn, have been unable to accurately analyze the air quality impacts of many traffic management strategies, including ramp metering (Guensler et al., 2001). If changes in vehicular travel activity due to ramp metering can be measured or modeled, EPA's MOVES model can be used to estimate changes in emissions.

Implementation Concerns

Agency Cost

Transportation and public works agencies at local, regional, and state levels use local, regional, state, and federal funds to implement ramp meters. Costs include the participation of traffic engineers and ITS specialists to implement traffic optimization plans, as well as technology costs such as fiber-optic networks to relay real-time traffic information.

Costs to the implementing agency include the ITS costs of model calibration, infrastructure implementation, and maintenance. More extensive ramp metering systems have a centralized control center. The cost of a particular ramp metering system varies widely according to the sophistication of the algorithm used to set the metering rate and the number of ramps included in the system (Pearson et al., 2003).

One example of capital costs for a proposed ramp metering system in the northern San Joaquin Valley of California in Stanislaus, California shows a range between $50,000 (where other interchange improvements were already being conducted) and $267,000 per ramp[79], which includes design, construction (including ramp improvements), installation, and technology (DKS Associates, 2008). On average, the cost to construct ramp meters was $155,000 per ramp for the 212 ramps in the system.

The Minnesota DOT also conducted a cost analysis of ramp meters. Ramp meters were estimated to cost approximately $3.2 million per year for the 430 ramp meters, or about $7,500 per year per ramp meter (Cambridge Systematics, 2001). This is an annualized cost, which enabled the study to compare cost to annualized benefits (including hours saved in travel time, reliability, fatalities, property damage and emissions). Their analysis suggested a cost-benefit ratio of 5:1.

Agency Implementation Concerns

Most ramp metering systems have been implemented by partnerships between state and regional/municipal agencies. The cases reviewed did not reveal any significant institutional barriers or inter-agency concerns.

---

[79] Originally reported in 2006 USD.

Social Concerns

Implementation of ramp metering is often initially opposed by the public because of increased queues at on-ramps (FHWA, 2006), undesirable levels of traffic diversion to surface streets, and increased emissions and fuel consumption at ramps (Pearson et al., 2003).

Ramp meters may also produce benefits for suburban motorists at the expense of those that live within areas that have ramp meters. This perception of inequity is based on the assumption that the suburban motorist lives outside a metered area and therefore is not delayed by ramp meters when entering a freeway (Nevada DOT, 2006).

Campaigns to educate the public on the benefits of ramp metering have helped public acceptance of ramp meters. Without these education campaigns, ramp meters may be viewed as costly and ineffective, which may lead to problems with receiving funding for ramp metering systems (Nevada DOT, 2006). In addition to this initial opposition, equity issues may arise due to the fact that ramp metering often benefits longer trips rather than shorter ones (Pearson et al., 2003).

Other Costs/Barriers

None identified.

Interactions with Other Strategies

- Ramp metering may positively interact with any strategy that benefits from shared use of advanced traffic management systems (ATMS), such as incident management systems and advanced traveler-information systems (Sorensen, 2008) because they complement other strategies that also require automated detection, surveillance, and control functions.

Unique Co-benefits

Co-benefits may include:

- travel time savings: Ramp metering is estimated to save almost 40 million hours of delay in the U.S. each year (Shrank and Lomax, 2009). It is estimated that if ramp meters were used on all highways, ramp meters could save about 100 million hours of delay annually in the U.S. (Shrank and Lomax, 2009);

- improved safety (Cambridge Systematics, 2001); and

- reduced travel time variability: The Minnesota DOT conducted an experiment that consisted of turning off the 430 ramp meters in the Minneapolis-St. Paul region for seven weeks in 2000. The results showed there are travel-time savings from operating the ramp meters, but the most dramatic change was the 26% increase in crashes when the meters were de-activated. There was also a 14% increase in the volume handled by the freeway with the meters on. Reducing collisions, increasing volume, and improving the reliability of service on the freeway all help maximize the return from the freeway investment (Cambridge Systematics, 2001).

U.S. Department of Transportation
Federal Highway Administration

Unique Negative Impacts

Ramp metering may create long waits to enter freeways, which can divert traffic to alternative routes. While this benefits freeway traffic, it can increase traffic on these alternative routes, which may have other negative effects.

Where in Use

Use is widespread, although in most places the impetus is more likely improvements in traffic flow than emissions reduction. Some specific examples include:

- Minnesota (Twin Cities) (Cambridge Systematics et al., (2001); Levinson, et al., 2006);
- Madison, WI (Kim et al., 2004);
- Denver, Colorado (Kim et al., 2004);
- Portland, Oregon (Ahn et al., 2007; Kim et al., 2004);
- Seattle, WA (Kim et al., 2004);
- Los Angeles, CA (Ahn et al., 2007; Sorensen, 2008);
- Seattle, WA (O'Brien, 2000); and
- Atlanta, GA (Guensler et al., 2001).

Recommendations for Further Research

Further research is required to determine if ramp metering is actually a viable GHG reduction strategy. Such research needs to consider the effects of idling and acceleration at ramp meters, reduced congestion, increased speed on highways, and any other effects of ramp meters.

References

Ahn, Soyoung; Bertini, Robert L.; Auffray, Benjamin; Ross, June H.; and Eshel, Oren (2007). Evaluating Benefits of Systemwide Adaptive Ramp-Metering Strategy in Portland, Oregon Transportation Research Record: Journal of the Transportation Research Board, No. 2012, Transportation Research Board of the National Academies, Washington, D.C., pp. 47–56.

Bogenberger, Klaus; Keller, Hartmut; and Ritchie, Stephen (2001). Adaptive Fuzzy Systems for Traffic Responsive and Coordinated Ramp Metering.
http://www.its.berkeley.edu/itsreview/ITSReviewonline/july2002/trb/00012.pdf.

Cambridge Systematics, Inc. et al. (2001). Twin Cities Ramp Meter Evaluation, Executive Summary. Minnesota Department of Transportation.
http://www.dot.state.mn.us/rampmeter/pdf/finalreport.pdf.

DKS Associates in association with PB Americas. (2008). Northern San Joaquin Valley Regional Ramp Metering and High Occupancy Vehicle (HOV) Lane Master Plan: Draft final Report. http://www.stancog.org/pdf/draft-final-report-ramp-metering-hov-master-plan.pdf.

Federal Highway Administration (FHWA) (2006). Ramp Management and Control Handbook. FHWA-HOP-06-083.
http://ops.fhwa.dot.gov/publications/ramp_mgmt_handbook/faqs/ramp_faqs.htm.

Guensler et al. (2001). Evaluation of Ramp Metering Impacts on Air Quality: The Atlanta I-75 Case Study: Final Report. Georgia Institute of Technology.

Kim, Gyejo; Lee, Soobbeom; and Choi, Keechoo. (2004). KSCE Journal of Civil Engineering. Vol. 8, No. 3, pp. 335-342.

Levinson, David and Zhang, Lei. (2006). Ramp meters on trial: Evidence from the Twin Cities metering holiday Transportation Research Part A. 40 (2006) 810–828.

Nevada DOT. (2006). HOV/Managed Lanes and Ramp Metering Manual.
http://www.nevadadot.com/reports_pubs/HOV/.

O'Brien, Amy (2000). New Ramp Metering Algorithm Improves Systemwide Travel Time, TR News, July-August 2000, Transportation Research Board.

Oregon Department of Transportation-Metropolitan Branch. (1982). I-5 North Freeway Ramp Metering, Portland, Oregon: Project Development-Operation.

Pearson, R. et al. (2003). Ramp Metering. Web document hosted by the Institute of Transportation Studies at the University of California at Berkeley and Caltrans.
http://www.calccit.org/itsdecision/serv_and_tech/Ramp_metering/ramp_metering_summary.htm.

Piotrowicz, Gary and Robinson, James (1995). Ramp Metering Status in North America. Prepared for the U.S. Department of Transportation Federal Highways Administration.
http://www.oregon.gov/ODOT/HWY/ITS/PDFs/BenefitsofITS/rampmeteringstatus1995.pdf.

Schrank, David and Lomax, Tim (2009). 2009 Urban Mobility Report. Texas Transportation Institute. http://tti.tamu.edu/documents/ums/mobility_report_2007_wappx.pdf.

Sorensen, Paul, Martin Wachs, Endy Y. Min, Aaron Kofner, Liisa Ecola, Mark Hanson, Allison Yoh, Thomas Light, and James Griffin. (2008). Moving Los Angeles: Short-Term Policy Options for Improving Transportation. RAND Corporation.
http://www.rand.org/pubs/monographs/2008/RAND_MG748.pdf.

U.S Department of Transportation- Intelligent Transportation Systems Technology Overview reviews components of Arterial Management Systems including Advanced Signal Systems and Adaptive Signal Controls: http://itsdeployment2.ed.ornl.gov/technology_overview/AM.asp

Incident Management

Policy: The Texas Transportation Institute (2009) estimated that traffic incidents account for nearly 60% of the traffic delay experienced in the 50 largest U.S. cities. Incident management programs use patrols or ITS to quickly detect and clear traffic incidents, thereby reducing delays and congestion and, in turn, reducing fuel consumption and CO_2 emissions.

Emissions Benefits and Costs: Studies of individual urban incident management programs across the U.S. show varying impacts on GHG emissions, with calculated reductions ranging from 2 $MTCO_2$ to 23 $MTCO_2$ per incident (compared to situations with no incident management). Costs are not well known and depend on the technology and approaches used. One report estimated a cost of $15 per $MTCO_2$,[80] which reflects the effects of reduced congestion and operating costs but not technology costs. It is unclear whether this accounts for induced demand.

Implementation Concerns: Incident management programs are generally acceptable given the time and fuel saving benefits they offer. They may require inter-agency coordination across jurisdictions and transportation facilities.

Background

The Texas Transportation Institute (2009) estimated that traffic incidents account for nearly 60% of the traffic delay experienced in the 50 largest U.S. cities. In 1998, incident-related congestion delay in the 10 most congested U.S. urban areas ranged from 218,000 to 1,295,000 person-hours. The additional fuel consumed during the same period in these areas because of incidents alone ranged from 214 to 1447 million liters (56.5 to 382.3 million gallons) (PB Farradyne, 2000). This translates to between 0.5 and 3.4 million $MTCO_2$.

Incident management is the process of quickly detecting and clearing incidents in order to reduce delays and congestion. Because incident management can reduce congestion, it may also reduce fuel consumption and GHG emissions. According to Schrank and Lomax (2009), there are a total of 272 incident management programs in the 439 U.S. urban areas. Incident management is one of five prominent types of operational treatments implemented to mitigate congestion. Other treatments include ramp metering (see the section on *Ramp Metering*), signal coordination (see the section on *Traffic Signal Optimization*), access management, and high-occupancy vehicle lanes (see the section on *Ridesharing and HOV Lanes*). It is estimated that these treatments collectively saved 308 million hours of driver delay caused by congestion in 2007; half of these savings are a direct result of incident management programs.

---

[80] All values consistent with 2009 USD unless noted.

U.S. Department of Transportation
Federal Highway Administration

Policy and Implementing Organization

Agencies may implement incident management efforts through roadway service patrols and/or Intelligent Transportation Systems (ITS). Service patrols tour congested and/or high incident freeway sections to identify incidents and/or disruptions in the traffic stream and minimize their duration, thereby restoring full capacity to the facility and reducing risks of secondary crashes to motorists and injury to responders (PB Farradyne, 2000). ITS infrastructure, including dynamic message signs, computer–aided dispatch, and closed-circuit television, are used to detect accidents and sometimes help avoid accidents. In 2004, 32% of freeway miles in the U.S. were monitored by video to detect incidents, and 45% were covered by service patrols (USDOT, 2007).[81]

The public sector or public-private partnerships run most of these programs. Incident management programs can be undertaken by local, state, or regional transportation agencies and involve the cooperation of law enforcement and emergency services. Transportation planning and programming agencies (e.g., MPOs) may also support incident management programs with funding. Some state DOTs are establishing incident response programs in collaboration with law enforcement agencies; DOTs and MPOs can work together with elected officials, police agencies, and city/county transportation agencies to undertake these programs and coordinate them among jurisdictions (PB Farradyne, 2000).

Effects

Target Group

Incident management programs affect traffic flow on highways and major arterials where they are present. These programs have implications for both passenger and freight traffic.

GHG Effects

While the bulk of incident management literature focuses on safety, congestion, and intelligent transportation systems, there is also research on the effects of incident management on fuel consumption and GHG emissions. Quantifiable benefits primarily include reduced incident clearance times (how long it takes to clear an accident), reduced crash frequency, and reduced delays. While many estimates have been made regarding the reduction of fuel consumption due to accident management programs, estimates on benefits for a per-incident basis vary greatly.

The effectiveness of a particular incident management program depends on numerous factors, including the number and type of incidents that occur in the region, the level of congestion that results, and the speed with which incidents can be cleared. Most studies do not explicitly account for induced demand and it is still unknown whether non-capacity expanding programs that improve travel time (e.g., ITS and operational strategies) actually induce demand (Neudorff,

---

[81] Video monitoring and service patrols are complementary and therefore not mutually exclusive. Thus, one cannot infer that 77% (32% + 45%) of freeway miles were monitored by video or served by service patrols.

U.S. Department of Transportation
Federal Highway Administration

2010).[82] Data from the following programs illustrate the variation in absolute and per-incident effects.

- The Maryland CHART incident management program was estimated to save 4.84 million gallons of fuel on 20,515 incident clearances in 2005 (NTIMC, 2006). This amounts to approximately 235 gallons of fuel and 2 metric tons of CO_2 saved per incident.

- Florida's Road Ranger program is estimated to save 1.7 million gallons of fuel per month (FL DOT, 2005; NTIMC, 2006), or 20.4 million gallons and 0.18 million metric tons of CO_2 per year. The service patrol program is part of an extensive Florida DOT Division IV ITS and freeway operations system. A key element of the ITS and freeway operation is the SMART (Systems Management for Advanced Roadway Technologies) SunGuide Transportation Management Center. The Road Ranger Program reportedly assisted with 79 out of 111 average daily recorded events in Broward County (FL DOT, 2005), or 28,835 events each year. The program is estimated to save about 707 gallons of fuel and 6.3 metric tons of CO_2 per incident.

- Other studies have shown much higher fuel savings per incident. For example, an analysis of the San Antonio, Texas TransGuide System estimated 2,600 gallons of fuel—or 23 metric tons of CO_2—saved per major incident (Henk et al., 1997). The TransGuide system also reduced overall accidents, including primary, secondary, and inclement weather accidents, by 41% (Henk et al., 1997).

Estimated Cost per Metric Ton of CO_2 Reduction

The costs of incident management programs and per incident costs are likely to vary. The Florida Road Ranger Program cost $2,500,000 in 2005, equivalent to $2,760,000 in 2009 USD, which is approximately $93 per incident. Based on fuel savings of 707 gallons and 6.3 metric tons of CO_2 per incident, this program costs approximately $15/MTCO$_2$. Note that this reflects operational costs only and does not include start-up technology or data management center costs, which can be high.

Key Assumptions and Uncertainties

Estimates of traffic delay reductions depend on prevailing traffic conditions (traffic volume, incident topology, and roadway characteristics) in the corridor or region where services are provided, and so are difficult to generalize across programs. In addition, data from these programs do not explicitly account for induced demand, which may reduce or negate the benefits.

---

[82] According to Neudorff, "Operational strategies and ITS, while improving travel times and reliability, do not explicitly increase roadway capacity. Because of this different "context," an argument can be made that the estimated offsets in cumulative GHG reductions resulting from induced demand are much too high. Additional research to better understand induced demand from improved operational efficiency and ITS is critical" (Neudorff, 2010).

Data and Tools
None identified.

Implementation Concerns

Agency Cost
Agencies' costs for incident management programs include service patrol (e.g., vehicles, staff) and ITS infrastructure (U.S. DOT, 2007). Certain technology costs can be high, such as fiber-optic networks to relay real-time traffic information. For example, in 2006 the Florida DOT District IV spent approximately $15.5[83] million on 55 CCTV cameras, 224 detectors, and 55 miles of in-ground fiber optic communications (USDOT, 2007). Costs of Transportation Management Centers depend on the design and size of the facilities. Annual costs of incident clearance programs can be as high as $21.3 million per year (2009 USD), such as the Los Angeles Metro Freeway Service Patrol program (RITA, 2006).

Agency Implementation Concerns
Incident management programs may require coordination across multiple jurisdictions (Johnson and Thomas, 2001). Programs may include a variety of actors and relationships across and within scales of government (e.g., municipal-municipal; municipal-state), which may complicate the implementation process.

Social Concerns
Evaluations of incident management programs have shown that the public is supportive of these programs (PB Farradyne, 2000). However, public relations campaigns are necessary to maintain the high levels of support needed to protect the program from budget cuts and improve relationships among partnering agencies (USDOT, 2001).

Transportation planning agencies have recognized the severity of travel time reliability problems and have been choosing operational strategies such as incident management programs that focus on mitigating non-recurring traffic congestion and improving reliability (Cambridge Systematics, et al. 2005). As transportation agencies devote more resources to non-recurring traffic congestion, the funding for incident management programs may increase.

Other Costs/Barriers
None identified.

Interactions with other Strategies
- Implementing incident management and speed reduction programs (which also rely on law enforcement) would positively interact with each other by lowering costs and resources for both policies and increasing the cost effectiveness.

---

[83] Equivalent to $16.6 million (2009 USD).

U.S. Department of Transportation
Federal Highway Administration

- By improving non-recurring delay, there is a chance that demand will increase; therefore, this strategy could potentially interact negatively with TDM programs. However, this strategy could also interact positively with certain TDM programs, as certain transit programs, e.g. bus rapid transit, may benefit from reduced non-recurring delay.

Unique Co-benefits

Co-benefits include increased safety and reduced delay, as found in the following studies.

- The CHART program in the Baltimore, Maryland/Washington, D.C. area expanded to more automated surveillance with lane sensors and video cameras on the region's freeway system. A majority of the benefits that are associated with this system result from a 5% (2 million vehicle-hours per year) decrease in delay associated with non-recurrent congestion (COMSIS Corporation, 1996).

- The Puget Sound Region of Washington State implemented a freeway service patrol in August of 2000 (Nee and Hallenbeck, 2001) A study in Seattle was conducted in which archived incident data from six months following implementation were compared to pre-implementation data from the same six month period during the previous year. This study revealed a decrease in emergency response time from 9 minutes to 5.8 minutes. Faster response time was estimated to reduce annual vehicle hour delay by 13,048 hours and result in a cost savings of nearly $240,000.

- The Texas Transportation Institute estimated that incident management programs could save 143.3 million hours of delay if incident management programs were implemented in all 429 urban areas in the U.S. (Schrank and Lomax, 2009). Benefit/cost ratios from the reduction in delay between 3:1 and 10:1 (savings in delay, fuel, emissions, safety compared to total cost of programs) are common for freeway service patrols (Fenno and Ogden, 1998).

Unique Negative Effects

None identified.

Where in Use

Slightly more than half of major urban areas have incident management programs, although in most places the impetus is more likely improvements in traffic flow than emissions reduction. Some specific examples include:

- Maryland (Farradyne, 2000);

- San Francisco Bay Area/Highway Service Patrol (Skabardonis, 1995; Farradyne, 2000);

- Maryland (COMSIS Corporation, 2006);

U.S. Department of Transportation
Federal Highway Administration

- Florida (http://www.transportation.org/sites/ntimc/docs/Benefits11-07-06.pdf);
- Arizona (Olmstead, 2001);
- Houston, TX (City of Houston, 2007);
- Portland, Oregon (Bertini, et al., 2005);
- Seattle, WA (Nee and Hallenbeck, 2001); and
- Minnesota/Highway Helper (MnDOT, 2002; Farradyne, 2000)

Recommendations for Further Research

There is very little fuel consumption/GHG research on other causes of non-recurring traffic congestion, such as work zones, weather, and special events. Since these causes of non-recurring congestion can be influenced by DOT/MPO action, it would be beneficial in a separate effort to study all causes of non-recurring congestion to determine the GHG effects.

References

Bertini, Robert, Rose, Michael, and El-Geneidy, Ahmed. (2005). Using Archived ITS Data Sources to Measure the Effectiveness of a Freeway Incident Response Program. Submitted for presentation 84th Annual Meeting of the Transportation Research Board January 9-13, 2005. http://tram.mcgill.ca/Research/Publications/CometTRB.pdf.

Cambridge Systematics, et al. (2005). Traffic Congestion and Reliability: Trends and Advanced Strategies for Congestion Mitigation. Prepared for Federal Highway Administration.

Chang, G., Liu, Y., Lin, P., Zou, N and Point-Du-Jour, J. (2003). Performance Evaluation of CHART: Coordinated Highways Action Response Team, Year 2002 (Final Report). November 2003. Available: http://www.chart.state.md.us/readingroom/readingroom.asp.

City of Houston. (2007).Houston SAFEclear Program Overview. Available: http://www.houstontx.gov/safeclear/index.html.

COMSIS Corporation. (2006). *CHART Incident Response Evaluation Final Report*. Silver Spring, MD.

Fenno, D. and Ogden, M. Freeway Service Patrols: A State of the Practice. Transportation Research Record No. 1634, Transportation Research Board. Washington, D.C. 1998.

Florida DOT. (2005). Annual Report: Smart SunGuide TMC. http://www.smartsunguide.com/PDF/Annual%20Report%2006_JAN_31%20FINAL.pdf.

Jacobson, L., et al. (1992). Incident Management Using Total Stations, Seattle, WA.

Johnson, Christine and Thomas, Edward. (2001). Regional Traffic Incident Management programs: Implementation Guide. FHWA-OP-01-002. Intelligent Transportation Systems, USDOT, Washington DC.

Henk, Russel H., et al. (1997). Before-and-After Analysis of the San Antonio TransGuide System. Paper presented at the 76th Annual Meeting of the Transportation Research Board, Washington, DC, January 1997.

Minnesota Department of Transportation. (2002). Highway Helper 2002 Summary Report.

National Traffic Incident Management Coalition. (2006). Benefits of Traffic Incident Management. http://www.transportation.org/sites/ntimc/docs/Benefits11-07-06.pdf.

Nee, J. and Hallenbeck, M. (2001). Evaluation of the Service Patrol Program in the Puget Sound Region. Report WA-RD 518.1. FHWA, U.S. Department of Transportation, 2001.

Neudorff, Louis G. (2010). "Moving Cooler" – An Operations and ITS Perspective. http://www.movingcooler.info/Library/Documents/Moving%20Cooler_ITS%20Perspective_Neudorff_Final_02202010.pdf.

Olmstead, T. (2001). Freeway management systems and motor vehicle crashes: a case study of Phoenix, Arizona, Accident Analysis and Prevention 33, 2001 pp. 433–447.

Farradyne, PB. November 2000. Traffic Incident Management Handbook. Prepared for the Federal Highway Administration Office of Travel Management. http://floridaapts.lctr.org/pdf/incident%20mgmt_handbook%20Nov00.pdf.

Pearson, R. (2003). Incident Management. Web document hosted by the Institute of Transportation Studies at the University of California at Berkeley and Caltrans http://www.calccit.org/itsdecision/serv_and_tech/Incident_management/incident_management_overview.html.

RITA. (2006). The Los Angeles County Metro budgeted $20.5 million for the 2005 service patrol program. http://www.itscosts.its.dot.gov/its/benecost.nsf/ID/ABC500E2613898FE852572CA004F43C1?OpenDocument&Query=State.

Schrank, David, and Lomax, Tim (2009). Texas Transportation Institute, Urban Mobility Report. The Texas A&M University System. http://www.sanangelompo.org/text_files/UMReport%202009%20WEB%20July%2009.pdf.

Skabardonis, A.; Noeimi, H. Petty, K.; Rydzewski, D.; Varaiya, P. and Al-Deek, H. (1995). Freeway Service Patrol Evaluation. PATH Research Report UCB-ITS-PRR-95-5 University of California, Berkeley.

USDOT (2001). Regional Traffic Incident Management Programs. FHWA-OP-01-002 http://ntl.bts.gov/lib/jpodocs/repts_te/13149.pdf.

USDOT (2007). Intelligent Transportation Systems for Traffic Incident Management. FHWA-JPO-07-001 http://ntl.bts.gov/lib/jpodocs/brochure/14288.htm.

Speed Limit Reduction and Enforcement

Policy: A vehicle's speed affects its fuel consumption and, consequently, its GHG emissions. The optimal speed for most motor vehicles with internal combustion engines is approximately 45-55 mph, and traveling at higher speeds quickly increases fuel use. This policy seeks to reduce vehicle speeds on highways and throughways (and thus reduce GHG emissions) by lowering and/or enforcing speed limits.

Emissions Benefits and Costs: Speed reduction programs are estimated to increase fuel efficiency (and hence reduce emissions) by 2-15% depending on the actual speed reductions achieved. Costs are approximately $9 to $12 per ton of CO_2 and consist mainly of enforcement costs.

Implementation Concerns: Motorists in most U.S. states are accustomed to speed limits of 65 mph or higher, coupled with a moderate margin for speeding. Official and public resistance across the country undermined national speed limit compliance and enforcement in past years and led to Congress's repeal of the 55 mph speed limit requirement in the 1995 National Highway Designation Act. Similar resistance could be a major obstacle to reinstating new, lower state or national speed limits.

Background

A vehicle's speed affects fuel consumption and GHG emissions due to air resistance and engine design. The optimal speed for most motor vehicles with internal combustion engines is approximately 45-55 mph, and traveling at higher speeds quickly increases fuel use (American Association of State Highway and Transportation Officials, 2008; Center for Clean Air Policy, 2004). However, most highways and throughways in the U.S. currently have speed limits of 65 mph or higher, significantly less efficient than the optimal range. Therefore, speed limit reduction and speed control programs have been suggested as a GHG mitigation strategy.

Policy and Implementing Organizations

The policy is to reduce highway speeds of 65 to 75 mph to 55 or 60 mph, with the aim of ultimately reducing actual driving speeds. This speed reduction could be implemented on a national, state, and/or highway level. Although legislatures control speed limits, law enforcement agencies are responsible for enforcing these limits, and state DOTs are responsible for changing speed limit signs. In addition, law enforcement agencies and state DOTs can provide information and analyses in support of effective speed management and can incorporate infrastructural and legislative features that discourage high speeds (Burbank, 2009).

Effects

Target Group

This policy targets all highway and throughway travel.

GHG Effects

Research shows that vehicles' fuel efficiency increases as speeds approach approximately 55 mph and then drops dramatically (FHWA, 2002), by between 1-2%, for each mile per hour (mph) traveled above 55 mph (Garcia, 1996; Center for Clean Air Policy, 2004).[84] GHG emissions increase inversely, declining until about 55mph and then increasing at higher speeds.

There is substantial theoretical and practical research showing that a reduction in speed reduces GHG emissions. However, few studies have examined the efficacy of speed reduction programs (i.e., how a particular campaign or program at a particular level of enforcement affects driver behavior and achieves GHG reductions). The effects of such programs will vary and depend on the actual reduction in speed that occurs from the program, the number of vehicle miles traveled (VMT) at reduced speeds, and the fuel economy of the vehicles affected by the program. The studies highlighted below, a mix of U.S. and European research efforts, are some of the few that have examined the relationship between reduced speeds and fuel savings or GHG reductions, in theory or in practice:

- A 1996 study estimated that enforcement of the then-current highway speed limits (55 mph and 65 mph) in Washington State would annually save about 105 million gallons of gasoline, which equals about 933,000 metric tons of CO_2 (Washington State Energy Office, 1996).

- In 2003, the New York State Greenhouse Gas Task Force estimated that the GHG emissions reduction potential for the entire state from fully enforcing existing speed limits from 1990 emissions levels would be 0.047 MMTC in 2010 and 0.070 MMTC in 2020 (Center for Clean Air Policy, 2003).[85]

- The National Research Council estimated that the former U.S. national 55 mph speed limit reduced national highway fuel consumption by about 2%, and that it probably saved 2,000 to 4,000 lives per year, due to lower fatality rates in highway crashes (Greene and Schafer, 2003).

- Based on average fuel economies and fleet mixes in the U.S., a study from the Center for Clean Air Policy (2004) uses per VMT rate of emissions to estimate that if average speeds were reduced from 65 to 55 mph (holding all other variables constant) covering an area of approximately 1 million daily VMT, CO_2 emissions would be reduced by almost 11,000 metric tons annually.

- The International Energy Agency (2005) examined the potential of a 55 mph motorway speed limit to reduce oil demand in the case of a sudden disruption in supply. It is estimated

---

[84] The Center for Clean Air Policy (2004) and Garcia (1996) estimate a reduction in efficiency of approximately 1% and 1.5%-2%, respectively, for each mile per hour over 55 mph.

[85] The estimates from this study are significantly smaller than estimates from other studies (e.g., Washington State). The reasons for these differences were not apparent from the reports.

U.S. Department of Transportation
Federal Highway Administration

that this measure would achieve a 3.3% reduction in transport fuel use in European countries.

- In Canada, greater enforcement of the current highway posted speed limits is estimated to reduce 85 million metric tons of CO_2 from 2001 to 2020 for Canada (Government of Alberta, 2006).

- In Rotterdam, The Netherlands, the speed limit was reduced from 120 kph (75 mph) to 80 kph (50 mph) on a 3.5 kilometer stretch of congested highway and enforced using a camera system and automatic fines. CO_2 emissions on this stretch were reduced by 15% (European Environmental Agency, 2008). The reduced speed limit along this stretch of highway also resulted in calmer traffic that reduced downstream bottlenecks and congestion.

- A 2008 study in Sweden found that lower car speed limits led to fewer circulating cars in an urban setting, thereby reducing greenhouse gas emissions. While the study did not quantify emission reductions, it cited prior research that estimated the reduction in cars on the road to be 10% in the long term if the average travel speed drops by 10% and 20% for a 33% drop in average travel speed (Delepierre, 2008).

Finally, the effects of reduced speeds on travel time and delays are unknown, but may also affect GHG emissions. Speed reduction in some places such as Rotterdam, Holland also reduced congestion and bottlenecks (European Environmental Agency, 2008), which could additionally reduce GHGs. In other places, reduced speeds could increase congestion or reduce capacity, which could increase GHGs. Alternatively, reductions in the speed of traffic can reduce total vehicle miles traveled by increasing travel time, resulting in a 2-5% reduction in vehicle travel in the initial years after implementation of a 10% decrease in speeds (CCAP, 2006). The limited research literature on this topic suggests that these effects have not been extensively studied and could vary significantly.

Estimated Cost per Metric Ton of CO_2 Reduction

Only a few of the aforementioned studies include the inherent fiscal responsibilities of these programs, most of which stem from increased enforcement. The New York Greenhouse Gas Task Force estimated a cost of $12[86] per metric ton of carbon for increased enforcement. The Government of Alberta estimated an annual cost of $90 million[87] for increased enforcement. This equals a cost of $9 per metric ton CO_2[88] if projected reductions are actually achieved; this cost would increase if compliance is less and therefore GHG reductions are less.

Key Assumptions and Uncertainties

The effect of speed reduction programs ultimately depends on whether and to what extent drivers reduce their speed in practice. This depends in part on how strongly the limits are enforced and on

---

[86] Costs are updated to 2009 USD.

[87] Costs are updated to 2009 USD; originally reported in 2008 CAD.

[88] Costs are updated to 2009 USD; originally reported in 2008 CAD.

driving culture, since in many areas it is customary for drivers to travel faster than posted limits. It additionally depends upon how reduced speeds affect congestion and VMT, which may positively or negatively affect emissions.

Data and Tools

Motor Vehicle Emissions Simulator (MOVES) is EPA's recommended tool for mobile source GHG emissions analysis. This model can be used to estimate the effects of speed changes on emissions. Many agencies interested in GHG emissions reduction strategies already use this model for other types of analysis, including transportation conformity for criteria pollutants.

The Comprehensive Modal Emissions Model (CMEM) is also sometimes used to determine the effects of speed on emissions. CMEM is a public domain model that can interface with a wide variety of transportation models (e.g., TRANSIM) and data sets (e.g., location, speed, and acceleration) in order to perform fuel consumption analysis (Barth and Boriboonsomsin, 2008).

Implementation Concerns

Agency Cost

The costs for speed programs are primarily from law enforcement and borne by law enforcement agencies. There are also small costs to transportation agencies for signs, data collection and other support costs. Some agencies (especially departments of motor vehicles) may also incur costs if they undertake public relations campaigns regarding the programs. The Government of Alberta estimated a total annual cost of $90 million[89] for increased enforcement of its reduced speed—a total of $1.8 billion over 20 years (Government of Alberta, 2008). Winkelman and Dierkers (2003) estimate a cost of $1.4 million for a GHG reduction of 0.117 MMtCE.

Agency Implementation Concerns

Transportation and law enforcement agencies often work together to implement speed management programs. These programs usually cross local jurisdiction and sometimes cross regional boundaries, which requires inter-agency coordination. Another concern is that speed limit reduction policies are only effective with motorist compliance, and therefore enforcement is critical.

Social Concerns

Speed limit reductions were historically set by each state and states may not be amenable to lowering their speed limits, particularly to a nationally-imposed level. Programs may also be unpopular if the public believes speed reduction will increase travel times.[90] Whether speed

---

[89] Costs are updated to 2009 USD; originally reported in 2008 CAD.

[90] In 1974, spurred by the energy crisis, Congress passed a law limiting the national speed limit to 55 mph to ensure efficient highway fuel consumption. From the time of implementation, however, this national speed limit had very low driver compliance. According to the New York Times, research in 1982 found that 83% of drivers on New York State Interstate highways exceeded 55 mph. Further, official and public opposition to

reduction programs actually increase travel time is unknown; in the study of speed reduction in Rotterdam, the program actually improved congestion downstream, presumably decreasing travel time (European Environmental Agency, 2008).

Other Costs/Barriers
None identified.

Interactions with Other Strategies
- Speed programs will almost certainly benefit from eco-driving programs, which inform the public about the benefits of better driving habits and the negative effects of poor driving habits, thereby encouraging buy-in for and compliance with speed programs.
- Speed programs may also be more cost effective when combined with incident management, because both strategies use similar service patrols and ITS.

Unique Co-benefits
- Reduced speeds may improve motorist safety. A 5% decrease in average speed on highways leads to approximately a 10% decrease in injury accidents and a 20% decrease in fatal accidents (OECD, 2004).
- Similarly, speed reduction may improve pedestrian and bicycle safety.[91]

Unique Negative Effects
- In 1996, EPA published research that forecast that national NOx emissions would increase at least 5% in the following scenario: urban speed limits remain unchanged and rural speed limits increase to 65 mph except that those states with limits below 65 mph before 1974 would maintain those lower limits (EPA, 1995.)
- Another potential negative impact is on freight performance, given that a key freight efficiency performance measure is average speed (Jones and Sedor, 2006).

---

the 55-m.p.h. limit was growing nationally, particularly for Interstates. Opponents in 1982 stressed that such highways were built to be safe at substantially higher speeds and noted that public and governmental pressure to save fuel had diminished as supplies increased and prices dipped. In addition, there was significant variability in states' willingness to fully enforce the speed limit. For example, Arizona, Idaho, Montana, Nevada and Utah have replaced stiff speeding penalties with nominal "energy wastage" fines of only $5 to $15 for those caught driving between 55 and the pre-1974 limits. Thus, in 1987, Congress passed the Surface Transportation and Uniform Relocation Assistance Act, which made the 55 mph speed limit only apply to urban highways. Later, in the National Highway System Designation Act of 1995 (Public Law 104 – 59), Congress fully repealed the original law, allowing states to have near total flexibility on speed limit decisions. This policy could be set at a national level again or individual states could pass their own legislation.

[91] Department for Transport (UK): http://www.dft.gov.uk/cyclingengland/site/wp-content/uploads/2011/01/a02_speed_reduction.pdf.

Where in Use

No U.S. state has endorsed a state-wide speed limit reduction, though a few areas have done analyses of the GHG savings that speed limit reductions might yield. The Washington State Climate Action Team recommended a state-wide speed limit reduction (Garcia, 1996). So far, the WS DOT has reduced speed limits on select roads. The UK offers one example of a nationwide speed reduction program.

Recommendations for Further Research

Further research is needed to determine the relationship between speed limit reductions and changes in travel time, congestion, and VMT to help assess the net effects of these programs.

References

American Association of State Highway and Transportation Officials (AASHTO) (2008, April). *Primer on Transportation and Climate Change.*

Barth, Mathew and Boriboonsomsin, Kanok. (2008). Real-world carbon dioxide impacts of traffic congestion. Transportation Research Record, (2058):163-171, 2008.

Burbank, Cindy (Feb. 2009). Special report 299: Reducing transportation greenhouse gas emissions and energy consumption: A research agenda.

Burbank, Cindy (Oct. 2009). NCHRP Project 20-24(59) - Strategies for Reducing the Impacts of Surface Transportation on Global Climate Change. A Synthesis of Policy Research and State and Local Mitigation strategies.

Center for Clean Air Policy (2006). Transportation Emissions Guidebook. http://www.ccap.org/safe/guidebook/guide_complete.html.

Center for Clean Air Policy (2004). "Urban Form and Climate Protection": http://www.ccap.org/Presentations/Winkelman%20TRB%202004%20(1-13-04).pdf.

David Suzuki Foundation (1998). Canadian Solutions: Practical and Affordable Steps to Fighting Climate Change, David Suzuki Foundation and the Pembina Institute for Appropriate Technology (www.davidsuzuki.org).

Delepierre, Camille (2008). Slowing Down? Why cities should decrease car speed and why they do not. Lund University Master's Thesis, Lund, Sweden.

Department for Transport (UK): http://www.dft.gov.uk/stellent/groups/dft_rdsafety/documents/page/dft_rdsafety_504682-03.hcsp#P105_10751.

Environmental Protection Agency (1996). "Information from the EPA Office of Mobile Sources: Emissions Impact of Elimination of the National 55 mph Speed Limit." http://www.epa.gov/oms/invntory/envspoms.txt.

European Environmental Agency (2008). Success Stories within the Road Transport Sector on Reducing Greenhouse Gas Emissions and Producing Ancillary Benefits. EEA Technical Report No. 2.

Federal Highway Administration (2002). "Transportation Air Quality - Selected Facts and Figures": http://www.fhwa.dot.gov/environment/aqfactbk/factbk13.htm.

Federal Highway Administration (FHWA) (2008). Speed Enforcement Program Guidelines. National Highway Traffic Safety Administration. www.nhtsa.dot.gov.

Garcia, Nicholas. (1996). Greenhouse Gas Mitigation Options for Washington State. Prepared for the Environmental Protection Agency by the Washington State Energy Office, www.muni.org/Departments/health/environment/Adobe%20Documents%20for%20ESD%20Site/WA_Action_Plan.pdf.

Government of Alberta Transportation (2008). Backgrounder: Highway Speeds and Greenhouse Gas Emissions Reductions. Air quality and Climate Change Briefing. http://www.transportation.alberta.ca/Content/docType57/Production/SpeedLimitsBrief.pdf.

Green, David and Schafer, Andreas (2003). Reducing Greenhouse Gas emissions from the U.S. Transportation Sector. Pew Center on global climate Change.

Grimes, Paul, "Practical Traveler: The 55-M.P.H. Speed Limit." *New York Times*, December 26, 1982, http://www.nytimes.com/1982/12/26/travel/practical-traveler-the-55-mph-speed-limit.html, accessed March 30, 2011.

International Energy Agency/OECD, Saving Oil in a Hurry (2005).

Jones, Crystal and Sedor, Joanne (2006). Improving the Reliability of Freight Travel. Public Roads. Volume 70, No. 1.

Natural Resources Canada (NRCAN): http://oee.nrcan.gc.ca/transportation/personal/driving/autosmartmaintenance.cfm?attr=8#defensive).

OECD and European Conference on Ministers of Transport (2004). Speed Management: Summary Document.

Schafer, Andreas. (2000). Carbon Dioxide Emissions from World Passenger Transport Reduction Options. Transportation Research Record 1738 paper number 00-1182.

United States Government Accountability Office (2008). GAO-09-153R; November 7, 2008. http://www.gao.gov/new.items/d09153r.pdf.

Victoria Transportation Policy Institute TDM Encyclopedia. http://www.vtpi.org/tdm/tdm59.htm.

Winkelman, Steven and Dierkers, Greg (2003). Reducing the Impacts of Transportation on Global Warming: Summary of the New York Greenhouse Gas Task Force Recommendations. Transportation Research Record No. 1842, Paper No. 03-4053.

Roundabouts

Policy: Traffic signals can increase stop-and-go driving, causing sudden acceleration and deceleration, congestion, and excess idling, all of which reduce fuel efficiency and increase GHG emissions. Roundabouts are alternatives to traffic signals. Roundabouts are circular road junctions in which traffic enters a continuous one-way stream around a central island. Such traffic routing can reduce vehicle idle times and improve traffic flow, thereby reducing fuel consumption and emissions.

Emissions Benefits and Costs: Substituting a roundabout for a conventional signalized or signed intersection may reduce fuel consumption and CO_2 emissions by vehicles traversing that roundabout. Estimates suggest reductions of 16% to 30% in fuel consumption and fewer emissions at roundabouts than conventional intersections. However, the net GHG effect of replacing intersections with roundabouts remains largely unknown because fuel-efficiency benefits may be reduced or negated by emissions from roundabout construction. The cost effectiveness of roundabout construction is also unknown.

Implementation Concerns: Roundabouts are expensive to implement and the public may resist them if the benefits are not recognized. In addition, current driver behavior in roundabouts in the U.S. is tentative, which affects overall performance and reduces capacity (NCHRP 572, 2007).

Background

Roundabouts are circular road junctions in which traffic enters a one-way stream around a central island.[92] Roundabouts are safer than traditional signalized and signed intersections because all traffic moves in the same direction and generally moves slowly and evenly. Roundabouts may also reduce GHG emissions by reducing vehicle idling and fast acceleration and deceleration that is typical of stop-and-go traffic at intersections (Mandavilli et al., 2008). However, fuel savings, if any, depend on the amount of traffic at a given intersection and the type of intersection that is being replaced (Isabrands et al., 2008). Net GHG savings also depend on the emissions from constructing the roundabout.

Policy and Implementing Organizations

State and local transportation agencies can use public money to replace signalized or signed intersections with roundabouts.

---

[92] Modern roundabouts were developed in the United Kingdom in the 1960s and address deficiencies in the original traffic circles that were in use in the United States since 1905. Those deficiencies included high-speed merging and priority for the merging traffic (which was reversed in modern roundabouts) that resulted in congestion.

Effects

Target Group

Roundabouts can be applied to any road intersection with traffic flow control needs. Roundabouts indirectly affect driver behavior.

GHG Effects

While much of the literature on roundabouts focuses on safety and traffic flow benefits, several studies from the U.S., Europe, and Australia assess the effects on fuel consumption and GHG emissions. These studies find that roundabouts can reduce fuel consumption and GHG emissions by 16-30% when replacing signalized and signed intersections. However, caution should be used when interpreting these figures. Israbands (2008) noted that model-based analyses of roundabouts can be flawed due to the use of outdated non-EPA (USEPA) emission factors to calculate findings. In addition, GHG production through roundabout construction is uncertain and could reduce or negate post-construction benefits. Therefore, the effectiveness of roundabouts as a GHG mitigation strategy is unknown.

A study from Northern Virginia examined 10 signalized intersections and estimated the effects on traffic delay and safety if these intersections had been constructed as roundabouts. Annual fuel savings were estimated to be more than 200,000 gallons in total from the ten roundabouts (20,000 gallons per roundabout per year on average, equivalent to 177 metric tons of CO_2). The annual average daily traffic on the 10 intersections ranged from 14,000 to 46,000 vehicles, with an average of 27,000 vehicles per intersection (Bergh et al., 2005).

Other studies examine the percentage of fuel savings and GHG emission reductions that occur when roundabouts replace conventional intersections. These studies have found that such emission reductions range from 16% to 59% at the site of the intersection being replaced,[93] with most studies reporting that the benefits include GHG emission reductions between 16% and 30% as compared to the emissions at the site of the original intersection.[94]

- The Barenkreuzung/Zollikofen project undertaken in Bern, Switzerland (population of 1 million) replaced the two most heavily traveled signalized intersections with roundabouts. The new roundabouts saw fuel savings and GHG emission reductions of about 17% as compared to the conditions at the site of the intersection that was replaced (European Academy of the Urban Environment, 2001).

---

[93] The percent decline in emissions and/or fuel consumption reflects improvements relative to the fuel consumption and GHG emissions that would occur while traveling through the original intersection. However, many studies do not explicitly state the spatial area that constitutes an intersection or the extent of the study areas and corridors.

[94] Note that a percentage in fuel savings tells us the percentage in CO_2 emissions savings, and vice versa, since the two are linearly related.

- In a study of six intersections (five in Kansas and one in Nevada) where roundabouts replaced signalized intersections, the average hourly CO_2 emissions were reduced by 16% in peak morning hours and 59% in evening hours as compared to the emissions at the site of the intersections that were replaced (Mandavilli et al., 2008).

- One study examined the effects of replacing a signalized intersection with a roundabout in Vaxjo, Sweden. A "car-following" method was used to directly observe and measure speed and acceleration before and after the roundabout was installed. The study found that fuel consumption declined by 28% (Várhelyi, 2002).

- Niittymaki and Hoglund (1999) compared fuel consumption between roundabouts and signalized intersections in Finland and found a reduction of 30% in fuel consumption for roundabouts at the intersection.

One study that tested emissions from vehicles directly found that the type of intersection being replaced, the amount of traffic, and the time of day affects whether roundabouts reduce fuel consumption (Zuger and Porchet, 2001). The study evaluated four locations in Switzerland with varying traffic density. While the roundabout that replaced a signalized intersection reduced emissions, the roundabouts that replaced non-signalized intersections did not decrease fuel consumption. Therefore, roundabouts may increase fuel consumption when previous smooth flow is replaced by the deceleration and acceleration of roundabouts.

Also, Kakooza et al. (2005) found that with lighter traffic, roundabouts have less waiting time (hence less stop and go traffic that causes increased fuel consumption) than un-signalized and signalized intersections in terms of easing congestion. However, with heavy traffic, signalized intersections may have less waiting time than roundabouts, due to the long queue time at the entrance of roundabouts in heavy traffic.

Note that none of these studies consider the emissions from roundabout construction. If emissions from capacity expansion projects are any indication,[95] these emissions could significantly reduce or even negate the benefits in fuel savings.

Estimated Cost per Unit of GHG Reduction

Recent roundabout projects in the United States have shown a wide range in reported construction costs. Assuming 2009 US dollars in the following examples, costs ranged from $13,000 for retrofitting an existing traffic circle into a roundabout, to $667,000 for replacing a traffic signal with a roundabout at the junction of two state highways, with an average cost of approximately $333,000 per roundabout (NCHRP, 1998; Federal Highway Administration, 2000). Roundabouts built by state agencies on state highways generally cost more because they can involve substantial

---

[95] Capacity expansion projects can produce several thousand tons of CO_2 per lane mile (Williams-Derry, 2007).

grading and drainage, as well as relatively long splitter islands and many curbs. These state-built roundabouts cost approximately $465,000 to $667,000 each (NCHRP, 1998).[96]

Key Assumptions and Uncertainties

To estimate the fuel savings from roundabouts, it is necessary to know the fuel consumption from the replaced intersection as well as the type of intersection being replaced. Fuel consumption at signalized intersections varies. Furthermore, roundabouts replace a variety of intersections, including signalized intersections, all-way stop controlled intersections, and yield-sign controlled intersections. The amount of traffic at these intersections is also a key factor.

In most studies, the emissions from roundabout construction, operations, and maintenance are not known, but this may be a significant source of GHG emissions and reduce or even negate the benefits of roundabouts. In addition, the operations and maintenance costs are uncertain, and, although they appear to be less than the same costs for signalized intersections, they must be included in order to have an accurate estimate of overall costs and cost effectiveness.

Data/Tools

SIDRA is a signalized and un-signalized intersection design and research aid that is often used in roundabout development projects.[97]

Implementation Issues

Agency Cost

Roundabout implementation poses several costs to transportation agencies, including construction costs, engineering and design fees, land acquisition, and maintenance costs. The reported costs of installing roundabouts have been shown to vary significantly from site to site (Federal Highway Administration, 2000). As noted earlier, costs ranged from $13,000 for retrofitting an existing traffic circle into a roundabout to $667,000 for replacing a traffic signal with a roundabout at the junction of two state highways, with an average cost of approximately $333,000 per roundabout (NCHRP, 1998; Federal Highway Administration, 2000). Roundabouts built by state agencies on state highways generally cost more because they can involve substantial grading and drainage, as well as relatively long splitter islands and many curbs. These state-built roundabouts cost approximately $465,000 to $667,000 each (NCHRP, 1998).[98]

A roundabout can be more expensive to construct than the two-way or all-way stop-controlled intersection alternatives, although it is difficult to compare the two since cost for roundabouts varies widely based on site-specific factors (Bergh et al., 2005).

---

[96] The costs have been converted to 2009 USD.

[97] http://www.sidrasolutions.com.

[98] The costs have been converted to 2009 USD.

Agency Implementation Concerns
No significant agency implementation concerns anticipated.

Social Concerns
Typically, roundabouts are supported for safety and congestion reasons, which are important to the public (GHG emissions are not usually a consideration). However, public acceptance can be one of the biggest challenges to implementing roundabouts because of misconceptions about roundabouts as outdated systems (Federal Highway Administration, 2000). Nevertheless, this can be mitigated with outreach efforts, and acceptance generally increases after roundabouts are implemented (NCHRP, 1998).

U.S. drivers are also less familiar with roundabouts than individuals in some other countries. This affects the overall capacity of roundabouts, because drivers are tentative when they drive through, reducing capacity and efficiency (NCHRP 572, 2007). This research identified that driver behavior was the largest factor in estimating roundabout performance.

Other Costs/Barriers
None identified.

Interactions with Other Strategies
- Using environmentally-friendly construction materials could decrease the emissions from roundabout construction, making roundabouts more effective in reducing GHG in an absolute sense and more cost effective. If the alternative material were lower cost than traditional materials, this would further improve cost effectiveness.
- When choosing a strategy to improve delay at intersections, agencies must decide between traffic signal optimization and roundabouts, which have similar goals but use different mechanisms that may not be complementary and that have different costs and benefits. Roundabouts are more expensive to implement but reduce more emissions (not counting construction emissions) (Barry, 2001). It was also found that the level of traffic (light to heavy) and volume variability of traffic (between merging traffic) may increase overall average stoppage time (Kakooza et al., 2005; Israbands et al., 2008).

Unique Co-benefits
A variety of co-benefits were found in the literature, including:

- improved motorist safety (FHWA, 2000; NCHRP Synthesis 264, 1998; Ahn et al., 2009);
- reduced congestion (FHWA, 2000; NCHRP Synthesis 264, 1998);
- greater pedestrian safety (FHWA, 2000);
- improved aesthetics/urban design (NCHRP Synthesis 264, 1998); and

- reduced noise at intersections since there is less acceleration from a full stop (FHWA, 2000).

Unique Negative Effects

None identified.

Where in Use

The majority of roundabouts are found in Europe and Australia. The U.S. has approximately 1,000 roundabouts in the eastern half of the country. A few examples include:

- Vermont (Redington, 2001);
- Florida (Kittleson and Associates, 2000); and
- Virginia (Bergh et al., 2005).

Recommendations for Further Research

There is a need for research on the energy consumption and GHG emissions from roundabout construction and other factors that may reduce their overall effectiveness in GHG mitigation. Such research is necessary to determine whether roundabouts are ultimately effective or ineffective. Project-level data could also be examined to understand costs and CO_2 reductions for roundabouts of various sizes.

References

Barry, C. (2001). Report on Roundabouts, January 2001, Downloaded from the website http://www.cccnh.org/cintroduction.htm.

Bergh, Casey; Retting, Richard, and Myers, Edward (2005). Continued Reliance on Traffic Signals: The Cost of Missed Opportunities to Improve Traffic Flow and Safety at Urban Intersections. Insurance Institute for Highway Safety. http://www.iihs.org/research/paper_pdfs/mf_1848.pdf.

Coelho, Margarida C., Farias, Tiago L. and Rouphail, Nagui M. (2006). *Transportation Research Part D: Transport and Environment, Volume 11, Issue 5, September 2006, Pages 333-34.*

European Academy of the Urban Environment (2001). Bern: Air quality management and traffic policy. Switzerland. Downloaded from the website http://www.eaue.de/winuwd/96.htm.

Federal Highway Administration (2000, June). *Roundabouts: An Informational Guide* FHWA-RD-00-67. *http://www.tfhrc.gov/safety/00068.htm.*

Hench, M. Quantitative Determining the Emissions Reduction benefits of the Replacement of signalized intersection by a roundabout.

Hydén C, Várhelyi A. (2000). The effects on safety, time consumption and environment of large scale use of roundabouts in an urban area: a case study. Accid Anal Prev. Jan 32(1):11-23.

Insurance Institute of Highway Safety, 28 July 2001. Status Report vol. 36(7).

Institute of Transportation Engineers. (2004). Signal timing practices and procedures: state of the practice. Washington, D.C.

Isebrands et al. (2008). Toolbox to Evaluate the Impacts of Roundabouts on a Corridor or Roadway Network. Minnesota Department of Transportation. http://www.lrrb.org/PDF/200824.pdf.

Kakooza, R., Luboobi, L.S., Mugisha, J.Y.T. (2005). Modeling traffic flow and management at unsignalized, signalized and roundabout road intersections. Journal of Mathematics and Statistics 1, 194–202.

Kittleson and Associates (2000). Roundabouts: An Informational Guide. US DOT. Publication No. FHWA-RD-00-067. http://www.tfhrc.gov/safety/00068.pdf.

Mandavilli, S. Rys, M. and Russell, E. (2008). Environmental Impact of Modern Roundabouts. International Journal of Industrial Ergonomics. 38, 135–142.

Mustafa, S., Mohammed, A., Vougias, S. (1993). Analysis of Pollutant Emissions and Concentrations at Urban Intersections. Institute of Transportation Engineers, Compendium of Technical Papers, Washington, D.C.

NCHRP Synthesis 264 (1998). Modern Roundabout Practice in the United States. Transportation Research Board. http://onlinepubs.trb.org/onlinepubs/nchrp/nchrp_syn_264.pdf.

NCHRP Report 572 (2007). Roundabouts in the United States, Transportation Research Board, http://onlinepubs.trb.org/onlinepubs/nchrp/nchrp_rpt_572.pdf.

New York Department of Transportation. https://www.nysdot.gov/main/roundabouts/files/Emissions_Reduction.pdf.

Niittymaki, J., Hoglund, P.G. (1999). Estimating vehicle emissions and air pollution related to driving patterns and traffic calming. In: Paper for the Conference on Urban Transport Systems, Lund.

Redington, Tony (2001). Modern Roundabouts, Global Warming, and Emissions Reductions: Status of Research and Opportunities for North America. Retrieved 11-27-2009 from: http://www.nh.gov/oep/resourcelibrary/referencelibrary/r/roundabouts/documents/vermontctrfpaper.pdf.

Schips, Norm, Hale, Terry and Rogers, Hal. NYSDOT Highway Design Manual. New York State Department of Transportation. Chapter 5.9.

Várhelyi, András (2002). The effects of small roundabouts on emissions and fuel consumption: a case study. Transportation Research Part D: Transport and Environment, Volume 7, Issue 1, January 2002, Pages 65-71.

Victoria Transportation Policy Institute TDM Encyclopedia.

Williams-Derry, C. (2007). *Increases in Greenhouse-gas Emissions from Highway-widening Projects.* Available at: http://www.sightline.org/research/energy/res_pubs/analysis-ghg-roads.

Zuger, Peter and Andre Porchet (2001). Roundabouts: Fuel Consumption, Emissions of Pollutants, Crossing Times. 1st Swiss Transport Research Conference. Monte Verita, Ascona.

Highway/Roadway Capacity Expansion

Policy: Expanding road capacity on congested highways can reduce traffic delays and improve mobility, potentially leading to reduced fuel consumption and GHG emissions. However, expansion may also increase demand, which would offset initial benefits and potentially lead to longer-term increases in fuel consumption and GHG emissions. Targeted capacity expansion seeks to reduce GHG by improving traffic flow on highways.

Emissions Benefits and Costs: Capacity expansion may not be an effective GHG mitigation strategy overall because GHG reductions from traffic flow improvements may be partly or totally offset by emissions from induced demand from new capacity. Capacity expansion costs approximately $4.05 to $7.1 million per lane mile,[99] but the costs per unit of reduction are not known given the uncertainties in GHG effects.

Implementation Concerns: While highway capacity expansion is often welcomed as a way to relieve congestion, it may not be effective in reducing GHG emissions in the long run, and is more expensive than other strategies.

Background

Expanding road capacity on congested highways can reduce traffic delay and improve mobility, and expansion projects are components of many urban congestion management programs. Since reduced traffic delay has been linked to reduced fuel consumption and increased GHG emissions, capacity expansion has been considered as a potential GHG mitigation strategy. Yet capacity expansion may simultaneously increase GHG emissions by ultimately generating more travel demand (increased trips and VMT) and increasing vehicle speeds (Niemeier, 2009; Cambridge Systematics, 2009; Cassady et al., 2004; Stathopoulos and Noland, 2003). Moreover, the process of capacity expansion itself can be a significant source of GHG emissions.

Policy and Implementing Organizations

Capacity expansion projects require coordination between several agencies. State DOTs, local governments (public works, etc.), and sometimes county governments can implement capacity expansion projects. Federal agencies (US DOT/FHWA) provide funding for certain capacity expansion projects. MPOs plan and allocate funding for these projects as well.

Effects

Target Group

This strategy targets the transportation network and indirectly affects users' travel demands and driving behaviors by providing more capacity for passenger and freight trips on highways.

---

[99] All values consistent with 2009 USD, unless noted.

GHG Effects

Capacity expansion projects are undertaken regularly, but the impact of capacity expansion on GHG emissions is controversial. The GHG effect of capacity expansion depends at least on the extent to which traffic flow improves (decreasing GHG emissions), the extent to which system-wide demand increases over time (increasing VMT and perhaps returning to previous congestion levels), and the energy and materials used for the construction project itself (increasing GHG emissions). There is little research evidence to support the conclusion that capacity expansion is effective as a GHG mitigation strategy given that some research has found that capacity expansion significantly induces demand and the emissions from this induced demand, in combination with construction, may outweigh any benefits.

GHG emissions from construction are generally estimated on a per-lane mile basis, though estimates vary depending on the nature of the construction. Some research suggested that building one lane-mile of roadway releases between 1,400 and 2,300 tons of CO_2, and long-term maintenance activities release between 3,100 and 5,200 tons of CO_2 (Williams-Derry, 2007 citing Graham, 2004). Using a more conservative estimate, Williams-Derry (2007) estimated that constructing one lane mile of highway and maintaining it for 50 years releases approximately 3,500 tons of CO_2. Using green construction materials could potentially reduce emissions from capacity construction.

When the emissions from construction are combined with travel effects, it is unclear whether CO_2 emissions are increased or decreased, given that research has found that highway capacity additions tend to ultimately increase VMT, particularly in growing areas where capacity increases attract further development (NCR, 1995; Cervero, 2003; Cervero and Hansen, 2002; NCHRP, 2005).

Project-level assessments show inconclusive results, in part because they do not usually account for embedded emissions (from production of materials such as asphalt) and life-cycle/cumulative emissions (e.g., from increased demand) (WSDOT, 2009). Some project-level studies have found the following results:

- Dutchess County, New York analyzed all strategies and projects within its long-range plan and TIP, concluding that the 2035 build scenario, which includes strategic roadway, transit, and non-motorized capacity expansion projects (versus the no-build scenario) would result in a 42 ton or 3.58% annual reduction in CO_2 based on reduction in VMT (PDCTC, 2007).

- The Environmental Defense Fund estimated that Maryland's proposed Intercounty Connector project, an east-west highway connecting the I-270 and I-95/U.S. 1 corridors in Maryland's National Capital Region, would increase CO_2 by 2.5 million metric tons of CO_2 per year, 0.1 million metric tons of CO_2 more than the no-build scenario, due to increased demand and fuel usage (Environmental Defense, 2005).

- Silva-Send (2009) estimated that a planned regional highway expansion project in the San Diego Region would save 8 million gallons of fuel (equivalent to 71,000 metric tons of CO_2) although it did not specify the time period for this saving. Moreover, it was unspecified

whether energy consumption from construction, maintenance, or induced demand where considered in the analysis.

- King County, Washington assessed the GHGs associated with replacing the 80-year old South Park Bridge, which was closed and demolished in 2010. This analysis compared the vehicle-generated GHG emissions that would occur with and without the bridge in place. The lack of a bridge was causing diversions of truck and car traffic onto already-crowded local streets. King County's analysis found that in 2006, over 14 fewer tons of greenhouse gases would be emitted each typical weekday with the bridge in place than without. By 2025, the GHG emissions reduction is projected to grow to over 34 tons per day—a growth rate of 4.84% in emissions savings each year (King County, 2010).

Estimated Cost per Metric Ton of CO_2 Reduction

This is not known given that capacity expansion projects' effects on CO_2 are highly uncertain and may increase or decrease.

Key Assumptions and Uncertainties

As suggested, a key factor in capacity expansion analysis is whether induced demand and complete life-cycle emissions (from construction, production of materials, maintenance, and other similar or related activities) are considered.

Data and Tools

None identified.

Implementation Concerns

Agency Cost

Capacity expansion is very expensive and costs are often underestimated (Litman, 2009). Estimates range from $4.05 million to $7.1 million per lane mile (2009 USD) for highway widening costs (Cox and Pisarski, 2003; Hartgen and Fields, 2006). Moreover, these projects are mostly in densely populated urban areas, and the costs do not always account for land acquisition, complex intersections, community mitigation, and delay costs during construction (Litman, 2009).

Agency Implementation Concerns

Highway projects that span multiple geographic jurisdictions may present challenges related to cooperation and collaboration across the various governmental agencies that may have a role in decision-making, permitting, planning and/or funding. Thus, lead project sponsor agencies and/or jurisdictions should pursue early and continuing efforts to engage concerned agencies at all governmental levels to facilitate the project development process and minimize conflicts and delays due to miscommunication or inadequate information sharing

Social Concerns

The public may favor capacity expansion projects because they appear to reduce congestion and improve travel time. However, there may be low acceptance of this policy as a strategy to mitigate GHG emissions.

Other Costs/Barriers
None identified.

Interactions with Other Strategies
Capacity expansion projects can reduce the effectiveness of many TDM strategies, especially those that encourage alternatives to SOV travel.

Unique Co-benefits
None identified.

Unique Negative Effects
In addition to GHG effects, capacity expansion may negatively affect the environment by changes in land use and increased development.

Where in Use
Many transportation agencies throughout the U.S. use capacity expansion as a congestion mitigation strategy.

Recommendations for Future Research
Since embedded emissions (production of materials) are not considered in most project-level analyses (WSDOT, 1999), estimates of capacity expansion are likely to be inaccurate and research should be undertaken to estimate the full effect.

References
Cambridge Systematics (2009). Moving Cooler: an Analysis of Transportation Strategies for Reducing Greenhouse Gas Emissions. Urban Land Institute.

Cassady, A., Dutzik, T., and Figdor, E. (2004). More Highways, More Pollution: Road-Building and Air Pollution in American's Cities, U.S. PIRG Education Fund.

Cervero R. and Hansen M. (2002). Induced Travel Demand and Induced Road Investment, a Simultaneous Equation Analysis. *Journal. of Transport Economics and Policy*, 36, 3, 469-490.

Cox, W. and Pisarski, A. (2004). *Blueprint 2030: Affordable Mobility and Access for All*. Georgians for Better Mobility.

Poughkeepsie-Dutchess County Transportation Council (PDCTC) (2007). *New Connections, Appendix B: Air Quality and Energy Analysis.* Poughkeepsie, NY.

Environmental Defense Fund (2005). *Maryland's Intercounty connector: Exacerbating Petroleum Dependence and Global Warming.*

Frey, C. and Rouphail, M. (2001). *Emissions Reductions Through Better Traffic Management: An Empirical Evaluation Based on On-Road Measurements*, North Carolina State University and North Carolina Dept. of Transportation. Available at: http://www.dot.state.nc.us/~research.

Graham, J.T. (2004). Hybrid Life-cycle Inventory for Road Construction and Use. *Journal of Construction and Management*, 130, 1, 43-49.

Hartgen, D. and Fields, M. G. (2006). *Building Roads to Reduce Congestion in America's Cities: How much and at what costs?* Reason Foundation.

King County (2010). South Park Bridge - Greenhouse Gas Analysis. http://www.kingcounty.gov/transportation/TIGERGrant/Environmental/GreenhouseGasAnalysis.aspx.

Litman, T. (2009). *Smart Congestion Reductions: Reevaluating the Role of Highway Expansion for Improving Urban Transportation.* Available at: http://www.vtpi.org/cong_releif.pdf.

National Research Council (NRC) (1995). *Expanding Metropolitan Highways, Implications for Air Quality and Energy Use*, Special Report 245.

NCHRP. 2005. *Predicting Air Quality Effects of Traffic-Flow Improvements*, Report 535.

Niemeier, Deb A. (2009). *Prioritization of Transportation Projects for Economic Stimulus with Respect to Greenhouse Gases: Final.* Prepared for the California Department of Transportation. Available at: http://transportationresearch.org/nl-ca/Shared%20Documents/UCD%20Presentations/Prioritization%20of%20Transportation%20Projects%20for%20Economic%20Stimulus%20with%20Respect%20to%20Greenhouse%20Gases.pdf.

Silva-Send, Nilmini (2009). *Reducing Greenhouse Gases from the On-Road Transportation in Dan Diego County, Executive Summary.* Energy Policy Initiatives Center, University of San Diego Law School.

Strand, et al. (2009). *Does Road Improvement Decrease Greenhouse Gas Emissions?*, Institute of Transport Economics of the Norwegian Centre for Transport Research. English summary at www.toi.no/getfile.php/Publikasjoner/T%D8I%20rapporter/2009/1027-2009/Sum-1027-2009.pdf.

USEPA (1998). *Traffic Flow Improvements, Transportation and Air Quality TCM Technical Overviews*, US Environmental Protection Agency.

Stathopoulos, Fotis G. and Noland, Robert (2003). *Induced Travel Demand and Emissions From Traffic Flow Improvement Projects.* Presented at the 82nd Annual Meeting of the Transportation Research Board.

Victoria Transport Policy Institute (2011). Road Pricing: Congestion Pricing, Value Pricing, Toll Roads and HOT Lanes. Available at: http://www.vtpi.org/tdm/tdm35.htm.

Williams-Derry, C. (2007). *Increases in Greenhouse-gas Emissions from Highway-widening Projects.* Available at: http://www.sightline.org/research/energy/res_pubs/analysis-ghg-roads.

WSDOT. (2009). *Washington State Department of Transportation Guidance for Project-level Greenhouse Gas and Climate Change Evaluations.*

Resurfacing Roads

Policy: Resurfacing rough roads reduces friction, thereby improving fuel efficiency and reducing GHG emissions.

Emission Benefits and Costs: Road resurfacing may not significantly decrease and may even increase GHG emissions because the process of resurfacing roads may produce significant CO_2, possibly more than the amount saved by the resulting smooth roads. Road resurfacing costs approximately $200,000 per lane mile; cost per metric ton of CO_2 is unknown and depends on traffic volume, fleet mix, and net GHG effects.

Implementation Concerns: Resurfacing may not be perceived as an effective GHG strategy given high costs and uncertain effects. However, resurfacing projects for safety and mobility reasons are otherwise well received.

Background

Road roughness results naturally from the gradual deterioration of road surfaces and/or the pavement structure. Not only do rougher roads reduce ride quality, they also reduce driver safety, increase vehicle wear and tear, and increase fuel consumption, which in turn increases GHG emissions (AASHTO, 2009). Road resurfacing has been suggested as a way to improve fuel consumption and reduce GHG emissions. Yet, it is unclear whether resurfacing roads actually reduces GHG emissions due to the energy-intensive process of resurfacing roads (Lepert and Brillet, 2009).

Policy and Implementing Organizations

The policy is to adequately maintain and resurface roads so that road conditions are at a lower roughness index. The expected effect is a reduction in GHG emissions due to improved fuel efficiency of vehicles from riding on smoother roads.

The policy could be implemented by any agency that is responsible for resurfacing and maintaining local, arterial, or highway roads, including state DOTs, local governments (e.g. departments of public works), and sometimes county governments (e.g., San Diego Association of Governments). MPOs, although not usually responsible for resurfacing roads, also could implement this strategy by planning for and allocating funds for road resurfacing projects. Many states already spend most of their transportation funds on road maintenance (Smart Growth, 2011).

Effects

Target Group

This strategy affects any road that has a high roughness index that can be lowered by resurfacing. This strategy does not require driver behavior changes, although road users benefit from lower vehicle operational costs and more comfortable travel.

GHG Effects

Several studies have examined the fuel economy differences between rough and smooth roads. These studies indicate fuel economy differences (and consequently GHG differences) between 1% and 10%, depending on the type of vehicle and the roughness of the roads considered.

- Studies have shown that reducing highway surface roughness through improved maintenance and using less flexible pavement surfaces such as concrete rather than asphalt can reduce fuel consumption by as much as 10% for heavy trucks, and by a smaller amount for lighter vehicles (BTE, 1996; TOI, 2009).

- In a Missouri study of vehicle performance on roads before and after paving, diesel dump trucks averaged 5.97 miles per gallon before repaving; after paving they averaged 6.11 miles per gallon. This 0.14 mpg difference is approximately a 2.4% improvement. A gasoline powered SUV averaged 21.30 mpg before and 21.47 mpg after, a 0.17 mpg improvement (slightly less than 1%). While these numbers are small per vehicle, with all the vehicle miles driven on the smoother roads of the Missouri Smooth Roads Initiative, these savings represent millions of gallons of fuel saved annually (MDOT, 2006).

- In one French study, researchers tested vehicles on stretches of roadway with various levels of road roughness to determine instantaneous fuel efficiency. They determined that, at constant speeds on level roads, fuel consumption increased 0.002 gallons per mile for a very rough road, whereas fuel consumption increased by 0.0007 gallons per mile for a slightly rough road. Thus, the researchers calculated that road surface characteristics can affect the fuel consumption of passenger cars by up to 7% (Du Plessis et al., 1990).[100]

- Another French study examined the difference in fuel economy for medium sized cars (weighing about 1,760 lbs) on roads with excellent evenness and fine macro-textures and on roads with poor evenness and exceptionally coarse texture. The fuel economy decreased from 34 miles per gallon to 32 miles per gallon (a decrease of approximately 6%) between the smooth and rough roads (Laganier and Lucas, 1990).[101]

The overall reduction in fuel consumption and GHG from a resurfacing effort depends additionally on the length of road resurfaced and the number and types of vehicles using the road. However, the fuel consumption effects are only part of the effects of resurfacing: CO_2 emissions are generated by road construction projects (including resurfacing) because of energy consumption, resource

---

[100] This study originally stated findings in metric units: for a road roughness index of 80, fuel consumption increased 5.28 mL/km, whereas for a road roughness of 15, fuel consumption increased by 1.72 mL/km.

[101] This study originally stated findings in metric units: an 800 kg vehicle that consumed about 7 L per 100 km on smooth roads and 0.4 L per 100 km (62.14 miles) on rough roads.

depletion, induced demand incited by the improvements, and increased travel speeds. Studies on the benefits of resurfacing rarely account for these emissions.

For example, one Australian study estimated a national GHG reduction from road resurfacing. The study suggested that from 1996 to 2015, a decrease in roughness by 40% could produce a cumulative reduction of CO_2 by 3.09 million metric tons on Australia's national highways (BTCE, 1996). However, emissions from the manufacture of construction materials and the equipment employed in road work were not taken into account and may be significant.

Indeed, a Norwegian report (Strand et al., 2009) suggested that road resurfacing does little to combat climate change. Road resurfacing emits 9.7 tons of CO_2 per lane mile for the construction alone.[102] It further found that road resurfacing increases CO_2 emissions because an improved quality of roads leads to higher speeds, especially if speeds increase to where the marginal effect of fuel efficiency on emissions is large (e.g., over 55 mph). Consequently, the net effects of road resurfacing on GHG emissions are not known and may be positive or negative.

Estimated Cost per Metric Ton of CO_2 Reduction

The research literature did not estimate costs per unit of reduction, and thus it is difficult to estimate this given uncertainties about the level and type of road traffic and, even more importantly, given that the emissions from resurfacing efforts themselves may be high but are often not reported. Agencies would most likely not resurface roads solely to reduce greenhouse gases.

Key Assumptions and Uncertainties

The effects of resurfaced roads vary depending on the number and type of vehicles on the roads and the changes in demand and use patterns that are induced by the improvements. The emissions and costs from road surfacing projects, which may be significant, depend on the type of material used, the transportation of that material to the project site, emissions from machinery, and other similar or related effects.

Data and Tools

The International Roughness Index (IRI) is an international standard developed by the World Bank used to measure pavement roughness. The index measures pavement roughness in terms of the number of inches per mile that a laser, mounted on a specialized van, jumps as it is driven. Specifically, the index is based on the "average rectified slope" (ARS), which is a filtered ratio of a standard vehicle's accumulated suspension motion (mm, inches, etc.) divided by the distance traveled by the vehicle during the measurement (km, mi, etc.). The lower the IRI number is, the smoother the road. It is based on a scale from zero for a true planar surface, increasing to about six (m/km) for moderately rough paved roads, to 12 (m/km) for extremely rough paved roads with

---

[102] This study originally stated findings in metric units: 12 tons of CO_2 for lane km.

U.S. Department of Transportation
Federal Highway Administration

potholes and patches, and up to about 20 (m/km) for extremely rough unpaved roads (BTCE, 1996).

Implementation Concerns

Agency Cost

Road resurfacing costs are generally high but vary depending on the road's current condition, location, and material. Road resurfacing can vary from less intensive preventative maintenance to more intensive reconstruction, and road resurfacing is generally part of agencies' maintenance and rehabilitations costs.

The average estimate for road reconstruction is $203,000 per lane mile[103] (Venner Consulting and Parsons Brinckerhoff, 2004). Missouri DOT estimated different costs per lane mile for interstate versus non-interstate: $128,700 per land mile for non-interstate roads and $319,000 per lane mile for interstate roads (Missouri DOT, 2008).

Major rehabilitation costs more than preventative maintenance. For example, preventative maintenance typically costs $56,750 to $114,500 per lane mile while reconstruction in urban areas is more expensive, sometimes exceeding $1,013,000 per lane mile[104] (Venner Consulting and Parsons Brinckerhoff, 2004).

Clearly, road resurfacing is expensive. Given the high costs and uncertainty about GHG benefits, road resurfacing is unlikely to be a strategy for reducing GHG, and instead would be undertaken for safety and mobility. Nevertheless, agencies would benefit from tools to measure GHG effects from road resurfacing projects and plans to allow full GHG analysis of agency activity.

Agency Implementation Concerns

This strategy may require some multi-level coordination between state DOTs and local governments, but overall, there are few agency implementation concerns.

Social Concerns

The public generally supports road maintenance for its safety and mobility benefits, and because of the public popularity of "fix it first" efforts. If GHG reductions occur, they would be considered secondary benefits and there would be little opposition. However, it would probably not be feasible (or effective) for agencies to attempt road resurfacing specifically to reduce GHG.

Other Costs/Barriers

No other costs/barriers were found.

---

[103] Costs are updated to 2009 USD.

[104] Costs are updated to 2009 USD.

Interactions with other Strategies
- If environmentally-friendly materials (whose production and use results in lower emissions than traditional materials) are used in the construction of roads, or improved practices such as recycling-in-place are used, the GHG reduction potential of road resurfacing may increase (see the section on "Green" Construction Materials for more information).
- In some instances, resurfacing roads may induce demand, which could act against travel demand management strategies.

Unique Co-benefits
- Road resurfacing may increase driver comfort and satisfaction.

Unique Negative Effects
Environmental impacts from road resurfacing include energy use that goes into construction materials, resource depletion, and energy and fuel used to resurface roads.

Highway resurfacing also generally causes some temporary disruption to traffic including reduced speed, increased delays, and increased crash risk due to altered road/ traffic conditions and the temporary absence of lane markings. Likewise, resurfacing causes temporary inconvenience, including noise, dust, and airborne particulate matter, to people living close to highways.

Where in Use
While road resurfacing occurs throughout the world, specific examples of road resurfacing efforts to reduce fuel consumption and GHG emissions have been researched in Missouri (Amos, 2006), Australia (BTCE, 1996), Norway (Strand et al., 2009), and France (Du Plessis et al., 1990).

Recommendations for Further Research
This strategy does not appear to be effective as a GHG reduction strategy since it does not seem to significantly reduce GHG emissions and is cost prohibitive. However, if agencies already have the responsibility of resurfacing roads, then GHG emission reductions calculations can be a useful tool for agencies that need to measure total GHG emissions from agency activity.

While more research is needed, it is clear that road resurfacing as a GHG reduction strategy has a very high cost/benefit ratio. As a secondary impact, agencies could calculate the GHG savings on a project-by-project basis. However, energy use from maintenance and excess delay caused from repaving must be considered. If road resurfacing materials are utilized that have lower energy use and environmental impacts, then there is potential for this strategy to mitigate congestion on a more efficient level.

References
AASHTO (2009). Rough Roads Ahead: Fix them now or pay for it later. http://roughroads.transportation.org/RoughRoads_FullReport.pdf.

Bureau of Transport and Communications Economics (BTCE) (1996). Transport and Greenhouse: Costs and options for reducing emissions, Report 94. Australian Government Publishing Service, Canberra, Australia.

Du Plessis, Hendrick w.; Visser, Alex T.; and Curayne, Peter C. (1990). Fuel Consumption of Vehicles Affected by Road-Surface Characteristics. In *Surface Characteristics of Roadways: International Research and Technologies,* Eds. W.E Meyer and J. Reichert. American Society for Testing and Materials, Philadelphia, pp. 480-496.

Fernández, P. C., and J. R. Long (1995). Grades and Other Load Effects on On-Road Emissions: An On-Board Analyzer Study. Presented at Fifth CRC On-Road Vehicle Emission Workshop, San Diego, Calif., Coordinating Research Council, Alpharetta, Ga.

Laganier, Robert and Lucas, Jean (1990). The Influence of Pavement Evenness and Macrotexture on Fuel Consumption. In *Surface Characteristics of Roadways: International Research and Technologies,* Eds. W.E Meyer and J. Reichert. American society for Testing and Materials, Philadelphia, pp. 454-459.

Lepert, Philippe and Brillet, Francois (2009). The overall effects of road works on global warming gas emissions. Transportation Research Part D 14, pp. 576–584.

Amos, Dave (2006). Pavement Smoothness and Fuel Efficiency: An Analysis of the Economic Dimensions of the Missouri Smooth Roads Initiative. Missouri DOT.

Missouri DOT (2008). Better Roads, Brighter Future. http://ep.modot.org/index.php?title=Category:Better_Roads%2C_Brighter_Future.

Park, S., and H. A. Rakha. (2006). Energy and Environmental Impacts of Roadway Grades. In Transportation Research Record: Journal of the Transportation Research Board, No. 1987, Transportation Research Board of the National Academies, Washington, D.C., pp. 148–160.

Patten, J.D. and Taylor, G.W. (2006). Effects of Pavement on Fuel Consumption. Centre for Surface Transportation Technology, Canada. http://www.mne.psu.edu/ifrtt/conferences/9thISHVWD/Presentations/02_3_Patten_Effects%20of%20Pavement%20Structure%20on%20Fuel%20Consumption%20.pdf.

Smart Growth America (2011). Recent Lessons from the Stimulus: Transportation Funding and Job Creation. http://www.smartgrowthamerica.org/documents/lessons-from-the-stimulus.pdf.

Strand, Arvid et al. (2009). Summary: Does Road Improvement Decrease Greenhouse Gas Emissions? Institute of Transport Economics.

Texas Transportation Institute (1994). Updated Fuel Consumption Estimates for Benefit-Cost Analysis of Transportation Alternatives.

Venner Consulting and Parsons Brinckerhoff (2004). NCHRP Project 25-25 (04) – Environmental Stewardship Practices, Procedures, and Policies for Highway Construction and Maintenance. http://onlinepubs.trb.org/onlinepubs/archive/NotesDocs/25-25(4)_FR.pdf.

Alternative Construction Materials

Policy: The majority of energy used for transportation construction comes from the production of pavement materials. Cement and asphalt production, in particular, are the largest sources of industrial process-related CO_2 emissions in the United States. Transportation agencies may use lower-energy alternatives instead of cement and asphalt to decrease GHG emissions.

Emissions Benefits and Costs: Emission reductions vary based on material but may be high and could be critical to the success of other strategies that depend on construction (e.g., capacity expansion). The costs of alternatives depend on the specific materials being considered, but in many cases costs may be small or negative since many materials are less expensive, or equivalent, to traditional materials.

Implementation Concerns: Barriers are low given the general cost effectiveness of these materials.

Background

The majority of energy used to produce transportation construction materials comes from the production of pavement materials (Huang et al., 2008; Zapata and Gambatese, 2005). Cement and asphalt production, in particular, are the largest sources of industrial process-related CO_2 emissions in the United States. In 2007, U.S. cement production emitted approximately 44.5 million metric tons of CO_2, slightly more than 0.7% of all CO_2 emissions for the year (Environmental Protection Agency, 2009).

Transportation agencies are beginning to use different materials in the construction process to decrease the adverse effects of construction on the environment. The most common alternative materials are forms of alternative pavement: *fly ash* instead of Portland cement, *warm- or cool-mix asphalts* instead of hot-mix, and *recycled road materials*:

- *Fly ash:* Fly ash, a by-product of coal combustion in coal-fired power plants, can be used to replace Portland cement in concrete. Portland cement is the binder material in traditional concrete and is associated with numerous adverse environmental effects including environmental degradation caused by mining the raw material, energy intensive procedures required for extracting and manufacturing the raw product, and CO_2 emissions during actual cement production.

- *Warm-mix asphalt:* New technologies have been developed to lower asphalt production and placement temperatures, and therefore the energy use, of hot-mix asphalt. These technologies are generally referred to as warm-mix asphalt (WMA), which uses substantially less energy and produces less CO_2 than traditional hot-mix asphalts in production and placement.

- *Recycled materials*: Recycled aggregates are produced from previously used road materials such as concrete and asphalt and are commonly used as a base layer for pavement construction in the U.S. However, there are concerns related to its durability (Tayabji, 2009). Recycled asphalt also shows promise in reducing production and construction energy requirements, though its benefits have not been adequately measured (Miller and Bahia, 2009). Used tires and shingles are sometimes incorporated into asphalt pavements. Other recycled content that can be used for road construction include recycled glass, wood ash, and paper mill residuals for use in concrete production (Naik and Moriconi, 2006).

Policy and Implementing Organizations

State DOTs and local governments (e.g., public works and county governments) can use construction materials that have lower energy requirements in their processing or application, are recycled, and/or have longer lives. Elected officials could pass legislation requiring recycled and environmentally friendlier materials in road construction and maintenance.

Effects

Target Group

This policy affects transportation construction and maintenance projects.

GHG Effects

There are several examples of agencies using alternative construction materials, but few research studies specifically calculate the energy savings and GHG reductions from using these types of construction materials. Moreover, energy savings and emissions reductions from the use of alternative materials vary depending on type of material, percent of recycled content, the scope of project, and other factors.

- *Fly ash cement:* Concrete used for highway construction consists of 10-15% Portland cement. Substituting coal fly ash for Portland cement can significantly reduce greenhouse gas emissions. Every ton of coal fly-ash substituted for Portland cement reduces life-cycle CO_2 emissions by almost one ton (Estakhri and Saylak, 2005). Using fly ash when replacing the current highway system over the next 20 years could yield 14 million metric tons of CO_2 savings by using 30% fly ash, and 24 million metric tons for 50% fly ash replacing cement in concrete.[105]

- *Warm-mix asphalt:* Using warm-mix asphalt instead of hot-mix asphalt reduces CO_2 emissions by 15-35% during production (D'Angelo, J. et al., 2008; Miller and Bahia, 2009). Half-warm mix and cold-mix asphalts can reduce energy consumption (and CO_2 emissions) by 50% (D'Angelo, J. et al. 2008). Another study reports that warm-mix asphalt results in a

---

[105] Based on 1.5 billion metric tons (Gt) of aggregates, 35 million metric tons (Mt) of asphalt, 48 Mt of cement, and 6 Mt of steel is in place in interstate highways (Sullivan, 2006).

reduction of CO_2 emission by about 9 kg (about 20 lbs) per ton of aggregate (Olard et al., 2008). If warm-mix asphalt replaced all traditional asphalt construction in the U.S. for future construction and maintenance projects, approximately 5 million tons of CO_2 would be reduced annually (D' Angelo et al., 2008).

- *Recycled materials:* Using recycled materials in roadbeds or for road surfacing may reduce GHG emissions. The 1.6 billion tons of cement produced annually requires about 2.5 billion tons of raw materials, usually limestone and clay. Replacing 50% of cement worldwide with other materials (e.g., fly ash, wood ash, etc.) would reduce CO_2 emissions by 800 million tons (Naik and Moniconi, 2006). This is equal to removing 25% of all automobiles from the world (Malhotra, 2004).

Estimated Cost per Unit of GHG Reduction

Some types of eco-friendly construction practices reduce costs (recycled materials), some are more expensive, and others are initially more expensive but reduce costs over the life-cycle of the project.

- *Fly ash:* Fly ash is an industrial waste and generally costs the same or less than Portland cement. The cost incurred is mainly that of transportation from the power plant to the construction site. As transportation costs increase, the cost benefits of fly ash are lessened (FHWA, 2003; U.S. EPA, 2005). When the price of fly ash concrete is equal to, or less than, the price of mixes with only Portland cement, fly ash concretes are given preference if technically appropriate under FHWA guidelines (Adams, 1988).

- *Warm-mix asphalt:* Across a series of case studies of warm-mix asphalt in several European countries, it was found that the cost of additives, asphalt plant modifications and related factors makes the cost of warm-mix asphalt greater than for hot mix asphalt, even when fuel savings are considered. Some officials believed that the likely longer life of warm-mix asphalt justified the higher cost (D'Angelo, J. et al., 2008).

- *Recycled materials*: The Michigan DOT (MDOT) has used recycled concrete aggregate (RCA) in road projects since 1983. MDOT has also found that incorporating RCA can reduce costs. For example, using recycled material aggregate for an interstate reconstruction project resulted in a total savings of $130,000[106] (FHWA, 2004).

Key Assumptions and Uncertainties

The type of alternative construction materials used varies energy use and GHG emissions, as do the materials they are replacing, so both materials' energy use (from production to construction) must be known in order to conduct an accurate analysis of the reduction. Transportation of construction materials should also be considered in analysis. For instance, due to the nature of fly ash as an

---

[106] Originally reported in 2004 USD.

industrial by-product, there is no primary cost to its production (because it would be produced regardless of the needs of transportation construction). However, the transportation of fly-ash from source to site has environmental and other costs and could offset the overall benefits.

Data and Tools
The EPA and FHWA have published guides about and on the use of fly ash. These are the guides:

- *Using Coal Ash in Highway Construction: A Guide to Benefits and Impacts (EPA, 2005).*
- *Fly Ash Facts for Highway Engineers (FHWA, 2003).*

Implementation Concerns
Agency Cost
The costs depend largely on the type of construction material used, which may affect transportation costs and production costs.

Agency Implementation Concerns
This strategy requires DOTs and other transportation agencies to use different materials and methods than they traditionally use, and this may necessitate a change in agency culture or the adoption of new policies to ensure the use of alternative materials. No significant inter-agency challenges are anticipated.

Social Concerns
If the alternative materials are not more expensive than traditional materials, then they should prove to be socially acceptable. To this end, several types of alternative road construction materials are currently used throughout the United States, indicating that this is the case.

Other Costs/Barriers
None identified.

Interactions with Other Strategies
- This strategy can be used alone but may benefit from strategies that increase the cost of fuel (thereby increasing the price of implementing conventional materials without increasing the price of alternative materials). This strategy would benefit, and may in fact be essential, for making other construction-based strategies effective in reducing GHG emissions, including capacity expansion, resurfacing roads, and roundabouts.

Unique Co-benefits
- Alternative materials may have less environmental impact in terms of resource depletion, water quality, air quality, and land use necessary for the construction of new materials and/or the disposal of material that would otherwise be waste.

- Alternative materials may also have a role in climate change adaptation, given that fly ash (FHWA, 2003) and warm-mix asphalt (D'Angelo, 2008) can be more adaptable to weather extremes.

Unique Negative Effect

None identified.

Where in Use

These materials (especially recycled asphalt) are in use throughout the United States and Europe. Examples of uses in the U.S. include:

- Texas: fly ash; warm-mix asphalt (Estakrhi and Saylak, 2005);

- California: sustainable concrete pavement (Tayabji et al., 2009); and

- Minnesota, Maryland, and Virginia: Recycled aggregate (www.fhwa.dot.gov/pavement/recycling/rca.cfm).

In the U.S. Recycled Materials Resource Center (RMRC) participants include the following transportation agencies: Caltrans, FDOT, Illinois DOT, Mass Highway, Michigan DOT, Mn/DOT, NHDOT, NJDOT, NYSDOT, NCDOT, Ohio DOT, PennDOT, TxDOT, and WisDOT. Participation indicates that DOTs are either interested in, or are currently using, recycled materials (www.recycledmaterials.org).

Recommendations for Future Research

There is generally a need for estimation tools to help agencies measure energy consumption from road works projects (Miller and Bahia, 2009). Life-cycle analysis (LCA) models capture lifetime costs but may neglect energy consumption and emissions; redefined LCA models are needed that include characteristics such as sustainability indicators and energy consumption (Huang et al., 2008).

References

AASHTO Center for Environmental Excellence. (2009). NCHRP 25-25 (4) Environmental Stewardship Practices, Procedures, and Policies for Highway Construction and Maintenance. http://environment.transportation.org/environmental_issues/construct_maint_prac/compendium/manual/detailed_toc.aspx.

Adams, T. H. (1988). Marketing of fly ash concrete. In MSU seminar: Fly ash applications to concrete (January), 1.10, 5.10. East Lansing: Michigan State University.

D'Angelo, J. et al. (2008). Warm Mix Asphalt: European Practice. Publication FHWA-PL-08-007. FHWA, U.S. Department of Transportation. http://international.fhwa.dot.gov/pubs/pl08007/pl08007.pdf.

Environmental Protection Agency. (2005, April). *Using Coal Ash in Highway Construction: A Guide to Benefits and Impacts.* Report EPA-530-K-05-002.

Environmental Protection Agency (2009, April). *Inventory of U.S. Greenhouse Gas Emissions and Sinks*, 1990-2007.

Estakhri, Cindy K. and Saylak, Donald. (2005). Reducing Greenhouse Gas Emissions in Texas with High-Volume Fly Ash Concrete. Transportation Research Record: Journal of the Transportation Research Board, No. 1941, Transportation Research Board of the National Academies, Washington, D.C. pp. 167–174.

Federal Highway Administration (2004). Recycled Concrete Study identifies current uses, best Practices. In *FOCUS*, April 2004. http://www.tfhrc.gov/FOCUS/apr04/01.htm.

Federal Highway Administration (2003). Fly Ash Facts for Highway Engineers. FHWA-IF-03-019. http://www.fhwa.dot.gov/pavement/recycling/fafacts.pdf.

Kahn Ribeiro, Suzana, Shigeki Kobayashi, Michel Beuthe, Jorge Gasca, David L. Greene, David S. Lee, Yasunori Muromachi, Peter J. Newton, Steven Plotkin, Daniel Sperling, Ron Wit, Peter J. Zhou (2007). Transportation and its Infrastructure. Climate Change 2007: Mitigation of Climate Change. Contribution of Working Group III to the Fourth Assessment Report of the Intergovernmental Panel on Climate Change, pp. 323 – 386.

Kapur, Amit; van Oss Hendrik G., Keoleian, Gregory; Kesler, Stephen E; and Kendall, Alissa. (2009). The contemporary cement cycle of the United States Journal of Material Cycles and Waste Management. Volume 11, Number 2.

Huang, et al. Development of Lifecycle Assessment Tool for Sustainable Construction of Asphalt Pavement. Eurobitume, (2008).

Lepert, Philippe and Brillet, Francois. (2009). The overall effects of road works on global warming gas emissions. Transportation Research Part D 14, pp. 576–584.

Miller, T and Bahia, H. (2009). Sustainable Asphalt Pavements: Technologies, Knowledge Gaps, and Opportunities. Prepared for the Modified Asphalt Research Center (MARC), University of Wisconsin.

Naik, T.R. and Moriconi, G. (2006). Environmentally-friendly durable concrete made with recycled materials from sustainable concrete construction.

Olard, F. et al. (2008). Low Energy Asphalt: New Half-Warm Mix Asphalt for Minimizing Impacts from Asphalt Plant Jo site. International ISAP Symposium on Asphalt Pavements and the Environment.

Rogers, Christopher D. F., Thomas, Andrew M., Jefferson, Ian, and Gaterell, Mark. Carbon Dioxide Emissions due to Highway Subgrade Improvements. (2009). Transportation Research Record:

Journal of the Transportation Research Board, No. 2104, Transportation Research Board of the National Academies, Washington, D.C., pp. 80–87.

Sullivan, Daniel E. (2006). Materials in Use in U.S. interstate Highways. US Department of the Interior U.S. Geological survey. http://pubs.usgs.gov/fs/2006/3127/2006-3127.pdf.

Tayabji, S., T. Van Dam, and K. Smith. (2009). Advanced Concrete Pavement Technology (ACPT) Program: A Status Report on Available Products. Report No. FHWA-HIF-09-005. http://www.fhwa.dot.gov/pavement/concrete/pubs/if09005/if09005.pdf.

van Oss, H.G. (2006). Cement: U.S. Geological Survey Mineral Commodity Summaries, p. 44-45.

Venta, G. J. (1999). Potential for Reduction of CO_2 Emissions in Canada Through Greater Use of Fly Ash in Concrete. Canada Centre for Mineral and Energy Technology/American Concrete Institute International Symposium on Concrete Technology and Sustainable Development, Vancouver, British Columbia, Canada.

Zapata, P., and Gambatese, J.A. (2005). Energy Consumption of Asphalt and Reinforced Concrete Pavement Materials and Construction. ASCE Journal of Infrastructure Systems, Vol. 11, No. 1, pp. 9-20.

7. Vehicle Improvement Strategies

Vehicle improvements strategies primarily seek to reduce GHGs by increasing the fuel efficiency of vehicles currently in use. These strategies are aimed at increasing the supply and demand for more fuel-efficient vehicles, and at increasing the fuel efficiency of currently-owned vehicles, either by improving the vehicles themselves or by improving how they are operated. This section also includes strategies that seek to reduce the fuel that is consumed when vehicles are used to perform other functions (e.g., waiting in vehicles for non-traffic reasons or heating or cooling the vehicle).[107]

Vehicle Strategies Reviewed in this Report

This review covers the following seven vehicle strategies:

- Feebates 175
- Scrappage Programs 181
- Tax Incentives for Cleaner Vehicles 189
- Heavy-Duty Vehicle Retrofits 195
- Eco-Driving Education and Training and Dynamic Eco-Driving 201
- Truck Stop Electrification and Auxiliary Power Units 207
- Anti-Idling Regulations and Campaigns 217

Government may use various market strategies to influence car-buying behavior. For example, feebates seek to increase the demand for fuel-efficient, conventional vehicles among those who are already in the market for a new vehicle. They do this by combining a tax on inefficient vehicles with a subsidy for efficient ones. Scrappage programs seek to increase the demand for new fuel-efficient vehicles among existing car owners who otherwise might not be in the market for a new vehicle. They provide financial incentives for vehicle owners to retire less fuel-efficient vehicles and replace them with more fuel efficient ones, earlier than they would otherwise have. Finally, tax incentives seek to increase the demand for alternative-technology vehicles such as hybrid electric or plug-in electric vehicles that have lower emissions by offering tax breaks to prospective buyers.

Heavy-duty vehicle retrofits and eco-driving strategies could each improve the fuel economy of currently owned vehicles. Heavy-duty vehicle retrofits change the aerodynamics of heavy trucks so that their fuel economy improves. Eco-driving strategies encourage drivers to adopt small changes in driving behaviors that can improve fuel economy.

---

[107] These are not vehicle efficiency improvements, per se, but they are more closely related to such strategies than to transportation demand management or system improvement strategies.

Finally, encouraging the use of truck stop electrification (TSE) or auxiliary power units (APUs), and implementing anti-idling regulations or campaigns both seek to reduce emissions from idling. TSEs and APUs are on and off-board technologies that allow long-haul truckers to use electricity to heat and cool their vehicles during resting hours, rather than idling their engines. Anti-idling regulations and campaigns seek to discourage private and commercial drivers from idling their vehicles unnecessarily (e.g., to warm vehicles in cold weather or while waiting for passengers or deliveries).[108]

Conclusions Regarding Vehicle Strategies

Strategies aimed at increasing vehicle efficiency through supply- and demand-side measures have generally been successful when implemented. Where they have yet to be implemented, modeling research still suggests they would be successful if well designed. Importantly, supply and demand side strategies are likely to interact positively: the fuel efficiency of vehicles in service may be increased by greater amounts and more quickly if both supply- and demand-side policies are implemented together, than if either is implemented alone.

The size of the effect on supply and demand of course varies depending upon the intensity of the incentives and penalties. Similarly, the resistance or acceptance of these strategies also depends on the extent to which they are voluntary or mandatory, on whether they are equitable, and on the size of burdens they place on manufacturers, consumers, and the public.

Strategies aimed at improving the performance of currently-owned vehicles have very different effects because they use different mechanisms to achieve those improvements. Retrofits involve physical modifications to trucks and are effective in reducing GHGs. They may face some opposition if made mandatory, even though they reduce operating costs for operators and payback periods are short. Eco-driving seeks to improve driving and vehicle maintenance behaviors that improve fuel economy and thus can be effective in reducing GHGs. Eco-driving campaigns and training programs face few barriers because eco-driving is voluntary, but the long-term effectiveness of driver instruction programs may be limited as drivers revert to prior habits. TSE and APU could reduce GHGs significantly among long-haul truckers and is well accepted because of the reduced costs to truck operators. Anti-idling regulations can also reduce emissions and have been implemented widely, largely out of concerns about air quality, but their effectiveness varies depending on enforcement.

The GHG effect of strategies that improve fuel economy of the fleet generally depends on the marginal improvement in fuel economy that the strategy achieves[109] and the number of vehicles or

---

[108] TSEs require improvements at truck stops, so they could also be thought of as transportation system management strategies. Anti-idling campaigns and regulations are not improvements to the vehicle, per se, but improvements to how the vehicle is operated, and are closely related to eco-driving.

[109] The marginal difference in fuel economy means different things for different strategies. For example, for heavy-duty vehicle retrofits, the concern is with the fuel economy of the same vehicle before and after the

consumers that are affected by the strategy. However, two phenomena counteract the GHG reductions from better fuel economy. First, an increase in fuel efficiency is vulnerable to induced demand. Research shows that the decrease in fuel costs from more fuel-efficient vehicles induces people to drive more. This effect has been estimated to be between 10-30% (UKERC, 2007), meaning, for example, that a 10% gain in fuel efficiency may result in a 1-3% increase in VMT. As this suggests, some vehicle strategies interact negatively with TDM strategies that seek to reduce VMT and encourage the use of other modes. Fuel taxes and road pricing—both of which make the act of driving more expensive—can be combined with other strategies that improve vehicle efficiency or transportation system efficiency to inhibit induced demand.

Second, some strategies like scrappage programs result in early vehicle replacement. By shortening the life of old vehicles, this results in higher rates of vehicle disposal and higher rates of manufacturing for new vehicles. Both processes may produce significant GHGs, and these life-cycle effects must be included when considering the effect of these supply and demand strategies. While scrappage programs are designed to retire vehicles early, other strategies may have this effect unintentionally or secondarily. For example, feebates are designed to influence the decisions of those who are already in the market for a new vehicle. However, if rebates are large enough, they could induce some consumers to buy new cars when they otherwise would not. This is less likely with other programs, but should still be considered as a possible side effect.

Finally, vehicle strategies have common co-benefits. In reducing the amount of fuel consumed, these strategies also reduce pollution, dependence on oil, and the amount travelers spend on fuel (except in the case of fuel taxes). Those that increase the supply or demand for new fuel-efficient vehicles also advance new vehicle technologies. Aside from fuel taxes, carbon taxes, and cap and trade programs, however, they do not reduce VMT and so they have no obvious effect on community livability, public health from increased use of non-motorized transportation, or congestion. Each strategy review also includes co-benefits other than those reported here that may be unique to the particular strategy.

References

UKERC (2007). *The Rebound Effect: An Assessment Of The Evidence For Economy-Wide Energy Savings From Improved Energy Efficiency,* The Technology And Policy Assessment Function Of The UK Energy Research Centre.

---

retrofit. For tax incentives for cleaner vehicles, the concern is with the fuel economy of, say, the hybrid vehicle that the consumer purchased because of the tax incentives versus the fuel economy of the sedan they would have otherwise purchased.

Feebates

Policy: Feebate programs create a monetary incentive for consumers to choose more efficient new vehicles by combining a tax on inefficient vehicles that have higher CO_2 emissions with a subsidy for efficient vehicles that have lower CO_2 emissions.

Emissions Benefits and Costs: The effect on emissions is sensitive to the design of the policy and cannot be generalized. The implementation costs are low because the tax on low fuel efficiency vehicles funds the subsidy for high fuel efficiency vehicles. Thus, the only cost is administering the program. Those costs are unknown because there are no existing feebate programs in operation, but costs should be within the range of other federal and state vehicle incentive (and "gas-guzzler" fee) programs.

Implementation Barriers: Several state legislatures have tried unsuccessfully to pass bills establishing feebate programs.

Background

Feebates create a monetary incentive for consumers to choose more efficient new vehicles by combining a tax on inefficient vehicles with a subsidy for efficient ones. Such programs can be implemented in a variety of ways. Feebates can be based either on either fuel efficiency (e.g., EPA estimates of miles per gallon or gallons per mile) or GHG emissions per mile. In the design of a feebate program, a *pivot point* is chosen where vehicles above that level of fuel efficiency receive a subsidy and those vehicles below it pay a penalty.[110] The penalty increases with declining fuel efficiency and the subsidy increases with increasing fuel efficiency. Feebate programs can also be differentiated within a vehicle class size, so as not to discriminate against larger vehicles, just less efficient ones within each vehicle class (Johnson, 2006).

Policy and Implementing Organizations

The policy of implementing a feebate program can be undertaken at the state or federal level. A federal program is likely to be more effective because there are problems with *leakage* at the state level. Leakage means that residents of a state with a feebate program might go to another state to purchase a low fuel efficiency vehicle while people from other states might travel to the state with the feebate program to purchase high fuel efficiency vehicles. As with other regulatory strategies, it seems unlikely that a state DOT or MPO would have much of a role in implementing a feebate strategy.

---

[110]It is also possible to create a schedule with a "zero-band" around the pivot point where vehicles are subject to neither penalties nor incentives.

Effects

Target Group

This strategy targets buyers of new cars and light-trucks in the region subject to the feebate program, although it is also possible to structure the program in such a way that manufacturers are assessed fees and receive rebates, which are then passed on to consumers (Greene et al., 2005). Technically, a feebate program could be implemented anywhere along the sales chain (i.e., at the manufacturer, dealer, or consumer level). However, Greene, et al. (2005) estimated that 90–95% of the response is due to manufacturers adopting new technology, making the manufacturer level economically efficient. At the same time, this option is likely infeasible for a state-level feebate program.[111]

GHG Effects

Feebates aim to encourage consumers to purchase more fuel efficient vehicles. Determining the effect on an individual consumer's choice and the savings achieved from that choice is difficult given the many assumptions that must be made.[112] The literature on U.S. and EU feebate programs is largely theoretical, due to a lack of implementation on which to build empirical evaluations. Unfortunately, the literature does not offer a consensus on design specifications or impacts. One can expect, however, that a program with steeper penalties (and, subsequently, steeper incentives) would result in more fuel savings. The quantitative studies that have considered the effectiveness of feebates have used models of consumer and manufacturer response to estimate the effects of applying incentives through a feebate program. There is consensus that manufacturers are likely to respond to a feebate by adding technology to their vehicle offerings, and that only a small portion of savings is likely to come from a shift to smaller vehicles (mix shifting).

Effects are sensitive to the design of the feebate program as well as to the level at which the program is implemented (i.e., state vs. federal). A lower pivot point and stricter feebate schedule will likely lead to greater reductions of GHGs. However, due to the multiplicity of design possibilities, the findings of specific studies as data points are presented below:

- Davis et al. (1995), using a combination of models to represent consumer choice and manufacturer response, found that 90% of the fuel savings achieved by a feebate results from technology adoption on the part of manufacturers. This reliance on technology development implies that feebate programs will be most effective if industry is given lead time to begin the process of adding technology to the vehicle offerings before the program is implemented.

---

[111] California is the possible exception due to its share of the new vehicle market and previous regulatory relationships with auto manufacturers.

[112] These assumptions include consumer preferences for vehicle attributes, consumer time discount rate, fuel prices, and technology costs.

- Koopman (1995) modeled a theoretical feebate program in Europe and estimated it could achieve average emissions of 0.64 lb/mi.[113] The study found an estimated emissions reduction of 10%, accompanied by a small net social cost (based on the loss in value to consumers from buying a different car than they otherwise would have purchased).

- Greene et al. (2005), estimated that a feebate of $500 per 0.01 gallons per mile (a fuel economy measure that is the inverse of mpg) would reduce new vehicle fuel consumption by 14% in the region in which it is implemented after manufacturers and consumers have had time to adjust to the new system of incentives.

- McManus (2007) assessed the potential effects of a continuous feebate schedule that offered up to $2500 in rebates for vehicles that emitted 100 grams of CO_2e per mile or less, imposed up to $2500 in fees for those that emitted more than 400 grams of CO_2 per mile, and a zero-band between 200 and 300 grams. The study found that new vehicle GHG emissions could be reduced by 17% under a feebates-only program, but could be reduced by as much as 33% when a feebate program is combined with GHG emissions standards in California. This result is likely an upper bound for a national program given the relatively high demand for fuel efficiency in California's vehicle market.

- de Haan et al. (2009) found that a shift to smaller engines was more likely than a shift to smaller vehicles, and that larger households, younger people, lower-income households, and households that prefer smaller cars were most likely to change purchasing behavior in response to a feebate program.

Estimated Cost per Metric Ton of CO_2 Reduction

As with fuel economy standards, most estimates of the net cost of feebates are negative because consumers will benefit from lower operating costs. As noted below, given uncertainty in fuel prices, consumers' time preferences, and their valuing of preferences for certain vehicle attributes, the cost could also be positive. The cost of administering the program is unknown because there are no functioning feebate programs, but it is unlikely to be much different than other vehicle incentive programs. It is possible that the costs of the program could be recovered from the fees paid, before rebates are awarded.

Key Assumptions and Uncertainties

The uncertainty in the estimates of costs and effectiveness results not from the quantity or quality of the studies, but from the lack of comparability across them. The specification and schedule of a feebate program must be very explicit, and each study in the literature essentially evaluates a different program (often under different assumptions). It is difficult, if not impossible, to definitively determine the GHG savings or cost of a generic feebate program; these programs and

---

[113] This was originally reported as 179g CO_2/km.

studies cannot be generalized to that degree. Furthermore, consumers' savings are based on estimated payback periods and assumptions about how consumers value fuel savings over time. This valuation will affect how consumers respond to the feebate incentives and influence the mix of vehicles and their relative efficiencies within each class of vehicle.

Data and Tools
None identified.

Implementation Concerns

Agency Cost
The cost of implementing a feebate program should be within the range of other federal and state vehicle incentive (and "gas-guzzler" fee) programs.

Agency Implementation Concerns
At the national level, much of the institutional capacity used to administer fuel economy programs could be used to administer a feebate program. However, at the state level, this capacity does not currently exist and would likely need to rely on EPA estimates of fuel economy—for Corporate Average Fuel Economy (CAFE) compliance—or regulatory agencies would need to develop new capacity for testing vehicle offerings. Furthermore, implementing a feebate at the manufacturer level, rather than at the consumer level, requires regulatory authority that a state agency may not currently have.

Social Concerns
A survey in the EU (de Haan et al., 2009) found that feebate systems were among the most accepted policy measures (scoring as high as purely informational measures, like energy labeling). This may not translate to similar consumer acceptance in the U.S. Several state legislatures have tried unsuccessfully to pass bills establishing feebate programs, suggesting that there may be challenges in such programs. Concerns have been voiced that feebates are another form of tax and that offering some consumers rebates while others are taxed is unfair (Johnson, 2006).

Manufacturers who sell more fuel-efficient cars and trucks will benefit more from a feebate program than firms that still primarily sell larger trucks. This may affect social acceptance, since it would seem to disadvantage domestic automakers that, until very recently, have largely focused on larger vehicles.

Other Costs/Barriers
Greene et al. (2005) estimated that the net social cost from a feebate program would be between $2 billion and $12 billion dollars, depending on design and payback periods. Society would have a net benefit under most designs coming from fuel savings from increased fuel economy; therefore, higher fuel prices make programs to increase fuel economy more cost effective for consumers. According to the study, at gas prices below $2/gallon, consumers can still save up to $2,000 over

the life of the vehicle for a well-designed program. This estimate is sensitive to assumptions about payback periods and long-term valuations.

Interactions with Other Strategies

- A feebate program may increase demand for and accelerate purchases of fuel-efficient vehicles, which would improve the practical adoption of more stringent fuel efficiency/emissions standards.

- Scrappage programs might be less costly to the public when combined with a feebate that provides an additional incentive to scrap older, less efficient vehicles.

- Similarly, the tax incentives for clean vehicles could be reduced if feebates were implemented simultaneously as an additional incentive.

Unique Co-benefits

- Increased technology penetration, since fuel-efficient vehicles use better technologies.

Unique Negative Effects

- It is possible that manufacturers would choose to reduce vehicle weight in order to maximize profit under a feebate program. Also, increasing the fuel efficiency of new vehicles reduces the cost of driving and may encourage drivers to travel more than they otherwise would have.

Where in use

There are currently no feebate programs in the U.S., though several states (notably Connecticut and Massachusetts) have tried to develop feebate programs recently. Washington, D.C. has implemented a variable vehicle registration fee which charges heavy vehicles more than the standard rate and hybrids half the standard rate. In Europe, France's environment ministry has proposed a feebate based on CO_2 emissions (Langer, 2005). The only actual automotive feebate program in the U.S., Canada, or the EU was implemented in Ontario, Canada, in 1991 and there are no quantitative studies of it, though it is generally considered to be ineffective because of its particular design: the program covered too few high-emitting vehicles (only 12% of all models) and used incentives of $75-100, too small to influence behavior (Langer, 2005). Canada's 2007 federal budget had funds allocated to create a prototype feebate program in Ottawa, but the program suffered from some of the same design flaws as the Ontario program In 2007, only ten vehicle models were subject to the feebate—making it easy for consumers to avoid the penalty. The program was also implemented immediately, without providing the lead-time that industry needs to adopt new technology. Leakage to the U.S. is also a concern for Canada, since the U.S. has no similar program in place. Canada is currently revising its feebate program and studying various implementation schemes (Banerjee, 2007).

Recommendations for Further Research

Further investigation of this policy requires detailed modeling of consumer demand for fuel economy that considers the impact of fuel prices, and the emerging national fuel economy program. While this is not a recommendation for near-term research, it would be a valuable area of future research. In the near term, it would be valuable to assess the social barriers to feebate programs that have been considered in the U.S.

References

Banerjee, Robin (2007). Deals on Wheels: An Analysis of the New Federal Auto Feebate, C.D. Howe Institute, No. 108, November.

Davis, W.B., Levine, M.D., Train, K., Duleep, K.G. (1995). Effects of Feebates on Vehicle Fuel Economy, Carbon Dioxide Emissions, and Consumer Surplus, DOE/PO-0031. Office of Policy, US Department of Energy.

de Haan, P.; Mueller, M. G. and Scholz, R. W. (2009). How much do incentives affect car purchases? Agent-based microsimulation of consumer choice of new cars - Part II: Forecasting effects of feebates based on energy-efficiency *Energy Policy, 37*, pp. 1083-1094.

Greene, D. L.; Patterson, P. D.; Singh, M. and Li, J. (2005). Feebates, rebates and gas-guzzler taxes: a study of incentives for increased fuel economy *Energy Policy, 33*, pp. 757–775.

Greene, David L. (2009). Feebates, footprints and highway safety, Transportation Research Part D, vol. 14, pp. 375 – 384.

Johnson, K. C. (2006). Feebates: an effective regulatory instrument for cost-constrained environmental policy *Energy Policy, 34*, pp. 3965-3976.

Koopman, G.J. (1995). Policies to reduce CO_2 emissions from cars in Europe: a partial equilibrium analysis. Journal of Transport Economics and Policy 30 (1), pp. 53-70.

Langer, T. (2005). Vehicle efficiency incentives: an update on feebates for states *American Council for an Energy-Efficient Economy*.

McManus, W. S. (2007). Economic analysis of feebates to reduce greenhouse gas emissions from light vehicles for California *University of Michigan Transportation Research Institute*.

Scrappage Programs

Policy: Scrappage programs provide financial incentives for vehicle owners to retire less fuel efficient vehicles and replace them with more fuel efficient ones, earlier than they would otherwise have.

Emissions Benefits and Costs: This is uncertain. While scrappage programs may reduce GHG emissions due to vehicle replacement, the life-cycle emissions from early vehicle replacement are typically unaccounted for but likely reduce this effect. The CARS program ("Cash for Clunkers") offers one estimate of $333/MTCO$_2$, without accounting for life-cycle emissions.

Implementation Barriers: Scrappage programs could be prohibitively expensive for smaller states or regions.

Background

Scrappage rates for modern vehicles are very low in the first three or four years of life, and most early scrappage is the result of traffic accidents. In later years, vehicles may additionally be scrapped due to reduced reliability, increasing maintenance costs, or preference for alternatives (Greenspan and Cohen, 1999).[114] Any program designed to accelerate this process may be referred to as a scrappage program.

Scrappage programs provide financial incentives for vehicle owners to retire older—and likely less efficient—vehicles earlier than they would under normal circumstances. They are often referred to as voluntary accelerated vehicle retirement (VAVR) programs, but have been known by more colorful names like vehicle scrappage, buy-back, or, more recently, "cash-for-clunkers" programs (Dill, 2004). The objective of VAVR programs is to substitute cleaner, more efficient vehicles for older, less efficient ones and accelerate the transformation in the population of registered vehicles that occurs naturally over longer periods of time.

Typically, these programs have been conceived to reduce conventional air pollutants like carbon monoxide, oxides of nitrogen, volatile organic compounds, and particulates emitted when vehicles burn fossil fuels.[115] For these purposes, an improvement can be realized by replacing an old vehicle essentially with any newer vehicle.[116] A number of countries have implemented age-based

---

[114] For reference, the median lifetime of a 1980 model year car is 12.5 years, while a 1990 model year car is 16.8 years. (*Transportation Energy Data Book (30th Edition)*, Table 3-11.)

[115] A long-running program in the San Francisco Bay area, the Bay Area Air Quality Management District's (BAAQMD) Vehicle Buy Back Program, scrapped over 10,000 vehicles between 1996 and 2004 for air quality purposes (Dill, 2004).

[116] Although cars have used emissions control technology for decades (catalytic converters, for example), these systems, too, degrade over time, causing older vehicles to emit much higher levels of conventional pollutants than new model vehicles (National Cooperative Highway Research Program, 1997).

scrappage programs. For example, France's program began in 2007. It initially limited retirement to cars over fifteen years old and then expanded it in 2008 to cars over 10 years old. A program in the UK ran from 2009 to 2010 and limited incentives to vehicles over 10 years of age (Lorentziadis and Vournas, 2011).

However, to reduce CO_2 emissions, "newness" alone is not enough: the replacement vehicle must have a better fuel economy than the replaced vehicle. Therefore, merely encouraging owners to replace *older* vehicles with *newer* ones may not have the desired effect unless the program explicitly accounts for the change in fuel economy between the old and new vehicles.

For example, the recent Car Allowance Rebate System (CARS), more commonly referred to as "cash-for-clunkers," targeted not just older cars, but older cars with an EPA estimated combined city/highway fuel economy of 18 mpg or less (NHTSA, 2009).[117] Moreover, it specified that the new vehicle must have a combined fuel economy of 22 mpg. The actual value of the incentive was determined by the fuel economy difference between the "clunker" and the new vehicle, and greater fuel economy increases received larger incentives. Incentives were between $3500 and $4500 in most cases, though some manufacturers and new car dealers provided additional financial incentives to spur demand during a challenging year for auto retailers. A detailed description of the incentive structure is available from a recent NHTSA report (NHTSA, 2009).

Other countries have also incorporated emissions requirements into their scrappage programs. In order to qualify for incentives, for example, France's program now requires that new cars purchased to replace old vehicles emit, at most, 160g/km (equivalent to 0.57 lbs/mile). Spain's program begin in 2008 and limits emissions to 149g/km (0.53 lbs/mi), and Portugal to 140g/km (0.50 lbs/mi) (European Automobile Manufacturers Association, 2010).

One subtle but important aspect of these programs is the life-cycle implications of scrapping vehicles earlier than would otherwise occur. That is, the emissions benefit gained from scrappage must also offset the emissions that are produced from the manufacture of the new vehicle. Kim et al. (2003) find that, for mid-size model year cars in 2000 and beyond, trading in a 7-14 year old vehicle minimizes conventional pollutants, but that a vehicle must be at least 18 years old to sufficiently offset the GHG emissions from prematurely manufacturing the new vehicle. Emissions may also result from the scrappage process. This has implications for VAVR programs, which may encourage much newer vehicle trade-ins, and indicates that policy makers should take a longer view of emissions than those originating at the tailpipe.

---

[117] A vehicle's estimated *combined* fuel efficiency is a weighted average of the highway (45%) and the city (55%) fuel efficiency estimates.

Policy and Implementing Organizations

A VAVR program could be implemented at the national level by a federal agency, like the 2009 CARS program ("cash-for-clunkers") or by DOTs and even MPOs, like the BAAQMD Vehicle Buy Back Program in the San Francisco Bay Area region.

Effects

Target Group

VAVR programs to improve GHG emissions target owners of cars with poor fuel economy. Typically, these inefficient vehicles are older, but a program specifically designed to reduce tailpipe CO_2 emissions could conceivably target relatively recent model year vehicles with poor fuel economy. For example, a program could provide incentives for people to trade-in very large SUVs (above 8,500 lbs) for either more efficient SUVs or passenger cars; this would have a greater CO_2 benefit than simply replacing an old fuel-efficient vehicle with a new one. However, if life-cycle CO_2 emissions are the measure of effectiveness, then only owners of still-older vehicles should be targeted (e.g., 18 years for the case of mid-size cars in model year 2000 and beyond).[118] In either case, the target group is current vehicle owners using inefficient vehicles. Additionally, there is some concern of "free ridership" in the program, where some portion of incentives go to vehicle owners who intend to retire their vehicles anyway.

GHG Effects

The literature evaluating specific VAVR programs to reduce GHG emissions is sparse, since VAVR programs are not typically used to achieve greenhouse gas reductions.[119] Moreover, there is currently no generally accepted approach to estimating GHG savings—with some studies using the fuel cycle (and tailpipe) emissions (NHTSA, 2009), others using a full life-cycle perspective (Kim et al., 2003, 2004), and others considering only the incremental increase in average fuel economy of new vehicles under a VAVR program (Sivak and Schoettle, 2009). Additionally, embedded within each of these estimates are important, but contentious, assumptions about VMT (for both the new vehicle and the vehicle it replaced), useful life, and the timing of new vehicle purchases in absence of the program.

Using survey data from the BAAQMD program and a pilot program in Southern California, Dill (2004) estimated that the average scrapped vehicle would have lasted between 1.8 and 3.2 more years in private ownership had the scrappage program not been in effect. However, as with other incentive-based programs, the specific effect on an individual's decision to replace a car could not be reliably quantified given the influence of other factors and variations in preferences. Moreover,

---

[118] In the recent CARS program, less than 10% of vehicles retired under the program were 18 years or older (NHTSA, 2009).

[119] Even the recent federal CARS program, which made some attempt to create incentives to improve fuel economy of new vehicles purchased under the program, was intended first as an economic stimulus program.

U.S. Department of Transportation
Federal Highway Administration

the per-person effects once a person has chosen to replace a vehicle are also difficult to quantify because this depends on the change in fuel economy between the retired vehicles and the new vehicles replacing them, as well as the driving behavior for each vehicle. For example, replacing a seldom-driven vehicle with a vehicle that will be driven extensively may lead to more CO_2 emissions despite the increase in fuel economy between the vehicles. The EPA has estimated a rebound effect of 10%, meaning that 10% of the fuel savings expected to result from an increase in economy is offset by additional fuel use.[120] Given this, it is not possible to calculate per-person CO_2 effects in a general sense. The CARS program conducted a survey to estimate the average additional length of time a retired vehicle would have been driven in the absence of the program. The average estimate was 2.87, which is within Dill's (2004) range of 1.8 to 3.2 years.

Estimating the aggregate emissions reduction from a VAVR program requires making many assumptions, which are described under "Key Assumptions." The NHTSA CARS report (2009) suggests that the recent "cash for clunkers" program led to changes that will reduce greenhouse gas emissions by nine million $MTCO_2e$ over the next 25 years. The estimated impact is based on the difference in fuel consumption between the population of registered vehicles with and without the CARS program, looking ahead to the end of the useful vehicle lives of the new cars and trucks purchased with the aid of the CARS incentives. The estimate attempts to account for both the changes in average fuel economy and travel behavior but *does not* consider life-cycle emissions, which would likely reduce this benefit.

The CARS program was of rather short duration, running only for two months in the summer of 2009, and it was the first of its kind at the national level in the U.S. Sivak and Schoettle (2009) found that the CARS program led to an increase in the average fuel economy of new vehicle purchases of 0.6 mpg and 0.7 mpg in July and August, respectively. They do not attempt to calculate GHG reductions resulting from this change in fuel economy, since this would require many additional assumptions about vehicle usage both before and after replacement.

Estimated Cost per Metric Ton of CO_2 Reduction

There are few estimates of the cost of these programs (and, aside from the literature described, almost no estimates of CO_2 reductions from their implementation). According to NHTSA, the CARS program saved CO_2 emissions at a cost of about $333 per $MTCO_2e$ reduced over the estimated 25-year life of the vehicles purchased under the program (NHTSA, 2009). This cost includes both the direct subsidies to new car buyers and the administrative costs of operating the program, but the emissions benefit does not take into account life-cycle emissions from vehicle replacement. Life-cycle emissions would reduce the emissions benefit and result in a higher cost per unit of reduction.

---

[120] Joint Technical Support Document: Rulemaking to Establish Light-Duty Vehicle Greenhouse Gas Emission Standards and Corporate Average Fuel Economy Standards, April 2010.

Key Assumptions and Uncertainties

Assumptions about driving behavior, both of the retired vehicle and the vehicle chosen to replace it, strongly influence estimates of program effectiveness.

- Since new vehicles are driven more than older ones (Small and Van Dender, 2005), providing incentives to substitute a new vehicle may lead to more driving, though it is more efficient.

- Substituting passenger cars for light trucks, and vice versa, further complicates estimation since light-trucks are driven more miles at each age and tend to have longer useful lives (NHTSA, 2009).

- There is also the possibility of a rebound effect, which would increase VMT as drivers pay less on a per-mile basis to travel with increased vehicle fuel economy.

- One must also make assumptions about when a vehicle would have been retired in the absence of the VAVR program in order to differentiate the effect of the program from the total effect of merely replacing the vehicle (which would have happened eventually anyway in most cases).

- Life-cycle effects (e.g., the emissions involved in producing and shipping new vehicles).

In sum, estimating the competing impacts of these factors requires making assumptions about complex issues about which there is little agreement in the literature.

Data and Tools
None identified.

Implementation Concerns
Agency Cost
Implementing agencies, at the federal, state, or local level, provide direct subsidies to consumers who are scrapping their vehicles (or possibly credits toward a new vehicle with the retirement of an older one). These costs can be significant, but can be estimated. These programs are likely to have a fixed amount of funding available for incentives, and scrappage programs can be designed to obtain the largest reductions with those resources. For example, the CARS program operated for nearly two months. This was based not on a fixed timeframe but on the time needed to distribute the budgeted incentives to new vehicle buyers. In addition to the cost of the actual incentives, programs will face administrative costs. The CARS program, for example, also resulted in $100 million in administrative costs. Major costs were transaction and voucher processing (estimated at $40 million for processing 20,000 dealer registrations and up to 250,000 transactions) and CARS information technology infrastructure (estimated at $30 million). These costs were high because of

the short time frame for development. Other significant costs included education and outreach ($3 million), and program staffing, travel, and general administration ($4.6 million) (NHTSA, 2009).

Agency Implementation Concerns

Smaller entities (MPOs or DOTs) implementing a VAVR program must ensure that retired vehicles are registered within their jurisdiction to avoid subsidizing another state or region's GHG reduction agenda. Nonetheless, there is no guarantee that vehicles purchased under this program will remain in that jurisdiction throughout their useful life.

Social Concerns

Although these programs have been acceptable for improving air quality, it is unknown whether a VAVR program to reduce GHG emissions would be similarly acceptable. However, the CARS program was well received in the sense that the initial funds allocated to the program were fully used within the first several days of the program—leading to an additional $2 billion supplement. A longer term program might meet with resistance as the total cost increases over time. Low-income individuals may benefit from this kind of strategy because they disproportionately own older vehicles. On the other hand, programs such as these could also drive up costs for used cars, which could adversely affect this population.

Other Costs/Barriers

None that are known to us at this time.

Interactions with Other Strategies

- Feebate programs may encourage consumers who retire vehicles under a VAVR program to purchase even more efficient vehicles than they would otherwise. Furthermore, a feebate program is likely to increase the fuel economy of most new vehicle offerings through technology adoption, thus making the average new vehicle purchased under a VAVR program even more efficient than the vehicle it replaces.

- Fuel taxes may also encourage consumers to purchase more efficient vehicles when retiring an older one under a VAVR program.

- Fuel efficiency/emissions standards will increase the average fuel efficiency of new vehicle offerings, increasing the fuel economy savings between a vehicle being retired and the new one being purchased to replace it.

Unique Co-benefits

- Accelerated vehicle emission and fuel economy technology penetration.

Negative Impacts

- By only considering the direct tailpipe CO_2 emissions of these retiring and replacement vehicles, one ignores the life-cycle implications of removing serviceable vehicles from the vehicle population and adding new ones. This may reduce direct tailpipe emissions, but raise the overall life-cycle emissions of vehicle ownership. Similarly, by encouraging owners to retire vehicles early, policies may contribute to increased road traffic and congestion as the new vehicles are driven more.

Where in Use

There are currently no retirement programs in the U.S. designed to reduce GHG emissions. Prior finite programs and the existing program in the San Francisco Bay Area likely achieve small reductions in GHG. There are a number of programs in Europe that include a GHG mitigation component (e.g., Germany, Spain, Portugal, Ireland, and France).

Recommendations for Further Research

There are no studies about the life-cycle carbon implications of VAVR programs and this warrants further study. There is an opportunity to use an experimental design with a scrappage program to quantify the causal relationship between the incentive and the consumer's vehicle scrappage decision making.

References

Dill, Jennifer. (2004). Estimating emissions reductions from accelerated vehicle retirement programs, Transportation Research Part D, 81, pp. 87-106.

European Automobile Manufacturers Association, (2010). *Fleet Renewal Schemes in 2010.* As of May 17, 2011:
http://www.acea.be/index.php/news/news_detail/fleet_renewal_schemes_soften_the_impact_of_the_recession/.

Greenspan, Alan and Darrel Cohen. (1999). Motor Vehicle Stocks, Scrappage, and Sales, *The Review of Economics and Statistics,* August, 81(3), pp. 369-383.

Kim, H.C., Keoleian, G.A., Grande, D.E., and Bean, J.C. (2003). Life Cycle Optimization of Automobile Replacement: Model and Application, *Environmental Science and Technology*, 27, 5407-5413.

Kim, H.C. Ross, M.H., Keoleian, G.A., (2004). Optimal fleet conversion policy from a life cycle perspective, *Transportation Research Part D*, 9, 229-249.

Lorentziadis, P.L, Vournas, S.G. (2011). "A quantitative model of accelerated vehicle-retirement induced by subsidy," *European Journal of Operational Research*, Volume 211, Issue 3, pp. 623-629.

National Cooperative Highway Research Program. (1997). Improving Transportation Data for Mobile Source Emission Estimates, Transportation Research Board.

National Highway Traffic Safety Administration (2009). Consumer Assistance to Recycle and Save Act of 2009: Report to Congress *U.S. Department of Transportation.*

Office of Transportation and Air Quality. (2009). Light-Duty automotive technology and fuel economy trends: 1975 through 2009 *United States Environmental Protection Agency.*

Sivak, Michael and Brandon Schoettle. (2009). The Effect of the "Cash for Clunkers" Program on the Overall Fuel Economy of Purchased New Vehicles, Transportation Research Institute, University of Michigan, UMTRI-2009-34.

Small, Kenneth A. and Van Dender, Kurt. (September, 2009). "The Effect of Improved Fuel Economy on Vehicle Miles Traveled: Estimating the Rebound Effect Using U.S. State Data, 1966-2001," Policy and Economics.

Tax Incentives for Cleaner Vehicles

Policy: Monetary incentives in the form of rebates or tax credits can be offered for purchases of alternatively fueled, fuel-efficient vehicles or vehicles with fuel-saving technologies.

Emissions Benefits and Costs: The effectiveness of incentives to change consumer behavior and reduce GHG emissions is unknown given that certain kinds of clean vehicles have only recently entered the market. Because consumers receive the incentive regardless of whether it influenced their decision to purchase a cleaner vehicle, the unit cost of reducing GHG is likely to be very high. Although much uncertainty makes it difficult to accurately quantify these costs, one high estimate was $3,700/MTCO$_2$.

Implementation Concerns: Incentives could be prohibitively expensive for smaller states or regions.

Background

The most effective fuel-saving technologies to reach mass market thus far are hybrid-electric vehicles (HEV) and clean diesel technology. While plug-in hybrid-electric vehicles (PHEV) and battery electric vehicles (BEV) may reach mass-market penetration by the end of the next decade, they currently represent a very small fraction of new vehicle sales. While these technologies have the potential to reduce fuel consumption, they are also typically more expensive due to the incremental cost of the technology over their conventional gasoline counterparts. Even at large production volumes, this incremental cost can be large and varies (depending on both technology and vehicle size) from about $2000 for a clean diesel engine in a passenger car to about $5100 for a hybrid drivetrain in a pickup truck or large SUV (Keefe et al., 2008). These technology costs are passed along to consumers who must weigh the additional cost against the potential fuel savings, and this deters many consumers from purchasing more advanced and efficient vehicles.

Governments, at both the state and federal level, have attempted to spur demand for these technologies by creating incentives to help reduce the initial cost to consumers.[121] The Energy Policy Act of 1992 established incentives for individuals and businesses to purchase clean-technology and alternatively fueled vehicles. In 2002, hybrid electric vehicles became eligible for these clean-fuel vehicle credits (Yacobucci, 2005). These were updated in 2005, when the Energy Policy Act of 2005 replaced an existing $2,000 tax deduction for all qualifying hybrid vehicles with a system of tax credits that applied to both hybrid and clean diesel vehicles, was sensitive to differences in fuel efficiency, and that terminated once a model achieved modest market

---

[121] This strategy encourages the adoption of particular fuel-efficient vehicle technologies (e.g., hybrids and clean diesel). Differential subsidies for more fuel-efficient vehicles (regardless of technology) also contribute to the uptake of these technologies, but they fall under feebate programs.

U.S. Department of Transportation
Federal Highway Administration

penetration (Lazzari, 2006).[122,123] Many states offer some tax incentive in addition to the federal tax credit. For example, Colorado offers tax credits of up to $6,000 (depending on the model), but this is currently the upper bound of credits provided by state programs.[124]

Policy and Implementing Organizations

Tax credit subsidies must be made by the same institutions and mechanisms that create and manage tax policies: the federal legislature at the federal level (e.g., as part of the 2005 Energy Bill) and state legislatures at the state level.[125] As stakeholders, transportation agencies and MPOs would likely be able to review and comment on legislative proposals to create tax incentive programs but would not directly administer them.

Effects

Target Group

These tax incentives target new car and light-truck buyers, specifically those buyers less inclined to pay a price premium for unfamiliar technology or who may not value fuel economy highly. However, the tax credits target all new car buyers, benefiting many early adopters and environmentally concerned buyers who would have bought these vehicles anyway.

GHG Effects

While tax incentives seek to influence individuals' purchasing behaviors, their per-person effects are unknown and essentially impossible to quantify, given the myriad of factors that affect an individual's purchasing decisions, as well as differences in priority among these factors.

Nevertheless, one can estimate the efficiency gains from these vehicles when they are purchased. Although effect size varies by vehicle class, on average a clean diesel engine (running on ultra low sulfur diesel fuel) has 25-30% better fuel economy and an HEV has 30-40% better fuel economy than their conventional gasoline counterparts (Keefe et al., 2008). Hybrid vehicles may provide still greater savings for urban drivers due to regenerative braking and greater use of the electric capacity at low speeds. EPA fuel economy estimates assume 55% of all driving is "city" driving,

---

[122] Credits varied from several hundred to several thousand dollars, but terminated when sales of the production model hit 60,000 units (Diamond, 2009).

[123] Qualifying clean diesel vehicles have only been available in the US since 2008. There are currently 12 diesel models and 31 HEV models that still qualify for some tax credit (www.fueleconomy.gov, accessed December 9, 2009).

[124] Many states have taken further measures to encourage hybrid ownership, such as providing access to high occupancy vehicle (HOV) lanes. While this may induce some consumers to purchase clean technology vehicles, the focus of this review is the impact of tax incentives on penetration rates.

[125] As a matter of application, consumers typically deduct the credit from the total tax burden in the year of purchase (for federal and some state tax credits), though others are structured so as to reduce the cost of the vehicle at the point of sale—such as a sales tax waiver (Gallagher and Muehlegger, 2008).

though this may be higher for drivers in dense urban areas. The greenhouse gas reductions from an individual vehicle depend on assumptions about both the VMT and on the fuel economy of the new vehicle that it replaces. For example, using the EPA assumptions that new vehicles travel 15,000 miles per year,[126] and assuming a conventional gasoline fuel efficiency of 30 mpg as a baseline, a hybrid version would save approximately 1.2 $MTCO_2$ in that year compared to the otherwise-equal conventional model. Analogously, calculating the overall impact of these policies requires making assumptions about the number of purchases that are affected by the incentive, the fuel economies of purchased vehicles, the fuel economies of vehicles that would have been purchased if the incentive were not in place, and the miles driven in each case.

It has only recently been possible to attempt evaluations of the tax incentive policies for clean technology vehicles, due to the limited availability of most qualifying vehicles. Such evaluations also need to separate the effects of the tax incentives from the effects of other factors (e.g., gasoline prices and social trends) on consumers' vehicle choices. Notably, because the current tax incentives were enacted during a period of exceptionally high gasoline price spikes, much of the observed increase in penetration was likely due to consumers' reaction to increased fuel prices rather than the tax incentives. For example, Gallagher and Muehlegger (2008) attributed only 6% of the increase in hybrid sales from 2000-2006 to tax incentives, compared to 27% and 33% of sales resulting from rising fuel prices and changing social preferences, respectively. Beresteanu and Li (2008) found that federal tax incentives were responsible for less than 4% of hybrid sales in 2005, but accounted for nearly 25% of new Prius sales in 2006. Diamond (2009) has supported Gallagher and Muehlegger (2008), suggesting that state financial incentives had little impact on the increase in hybrid vehicle sales over that time period. Diamond (2009) also found that tax incentives that affected the vehicle price at the point of purchase had a greater impact than did credits or rebates, which took longer to realize.

According to the Department of Energy's Alternative Fuels and Advanced Vehicles Data Center, domestic new hybrid vehicle sales increased from 210,000 units in 2005 to 274,000 units in 2010, peaking at over 350,000 units in 2007.[127] Vehicle offerings also expanded from 8 to 29 models in that 6-year time period. However, from 2000 to 2006 only 660,000 hybrids were sold. Gallagher and Muehlegger (2008) have implied that the tax incentives were responsible for the sale of less than 40,000 new hybrids during that time. Beresteanu and Li (2008) implied that approximately 25,000 units of the Toyota Prius were sold in 2006 as a result of the tax credits. The credits for the Prius have since expired, but it still leads all other hybrid models in sales.

One must make assumptions about the new vehicles that these hybrids would have replaced in order to estimate CO_2. One reasonable solution is to assume that the new vehicle sales replaced by

---

[126] www.fueleconomy.gov.

[127] EPA updates this information annually on its website at http://www.epa.gov/otaq/fetrends.htm.

the hybrid sales would have had the average fuel economy of the new vehicles sold that year. For simplicity, let us assume that the 40,000 new hybrids were purchased uniformly in each year between 2000 and 2006. The average fuel economy of new vehicles in this period was 29.4 miles per gallon, and the average miles driven for new vehicles is 15,000).[128] If hybrids have an average fuel economy of 50 mpg, then the incentive would have saved approximately 8.5 million gallons of gasoline and reduced approximately 75,000 metric tons of CO_2.

Estimated Cost per Metric Ton of CO_2 Reduction

Given the small percentage of hybrid sales growth that is likely attributable to tax incentive programs, the costs per metric ton reduction are likely to be very high. Estimates of this value are sensitive to assumptions (as well as cost accounting—many consumers took advantage of both federal and state tax incentives). Using the conventional gasoline model for comparison and EPA estimates of travel behavior, Diamond (2009) has estimated costs as high as $3,700/$MTCO_2$ reduction in some cases. The tax incentive typically varies by vehicle, so there is likely to be a wide range of cost effectiveness varying by both vehicle and state of purchase.

Key Assumptions and Uncertainties

The studies that have considered the effectiveness of tax incentive programs on clean vehicle technology penetration have concluded that fuel price is a key uncertainty, as are simultaneous incentives (either tax incentives at both the state and federal level or non-monetary incentives like HOV access) and evolving social preferences. To the extent that incentives encourage owners to retire older cars sooner than they would have otherwise, the life-cycle emissions from faster turnover may reduce GHG emissions. However, these effects are not well understood or included in most studies.

Data and Tools

Data about domestic clean vehicle technology sales is available from Department of Energy's Alternative Fuels and Advanced Vehicles Data Center.[129] Federal tax incentive programs are regularly updated at www.fueleconomy.gov.

Implementation Concerns

Agency Cost

The federal tax incentive program phases out based on sales volume (and over time) for popular vehicles so as to avoid unnecessary subsidies for vehicles that would be purchased anyway. However, states structure their programs differently and may bear large costs, either directly or as opportunity costs, on forfeited sales tax revenue.

---

[128] www.fueleconomy.gov.

[129] www.afdc.energy.gov/afdc.

Agency Implementation Concerns

There are no specific agency implementation concerns associated with tax incentives.

Social Concerns

These types of incentives have existed at the federal level for about a decade, and other federal programs have supported public-private partnerships to advance similar technologies. Given the additional incentives at the state level, social acceptability appears to be high.

Other Costs/Barriers

None that are known to us at this time.

Interactions with other Strategies

- Eco-driving education may change social preferences for vehicles employing fuel-efficient technologies.

- Feebate programs may create additional incentives to purchase clean-technology vehicles, which are likely to be among the more fuel-efficient vehicles in each class.

- Low Carbon Fuel Standards will act in concert with incentives for clean-technology and alternatively-fueled vehicles by encouraging wider availability of alternative fuels and developing alternative fuel infrastructure.

- Fuel taxes may negate the need for tax incentives to purchase clean-technology, high-efficiency vehicles since their market share may increase as a response to high conventional fuel prices.

Unique Co-benefits

- Greater technological development and technology penetration.

Unique Negative Effects

- There is an opportunity cost associated with creating this incentive program, given that research suggests it is somewhat inefficient. There is also a potential increase in VMT from the rebound effect, which would diminish the effectiveness of increasing fuel economy through these incentives.

Where in Use

There is a federal tax credit program for clean vehicle technology, as well as several state-level tax incentive programs typically targeting HEVs.

Recommendations for Further Research

There is a need for a causal analysis of the influence of incentives on demand for clean vehicles. State DOTs and MPOs can sponsor research and help to design research and data collection to this end.

References

Beresteanu, Arie and Shanjun Li (2008). Gasoline Prices, Government Support, and the Demand for Hybrid Vehicles in the U.S., Duke University.

Bureau of Transportation Statistics (2009). *National Transportation Statistics 2009*. US Department of Transportation.

Department of Energy Alternative Fuels and Advanced Vehicles Data Center URL: http://www.afdc.energy.gov/afdc. Accessed October 13, 2010.

Diamond, David (2009). "The Impact of Government Incentives for Hybrid-Electric Vehicles: Evidence From the US States", Energy Policy, 37, pp. 972-983.

Gallagher, Kelly Sims and Erich Muelegger (2008). "Giving Green to Get Green: Incentives and Consumer Adoption of Hybrid Vehicle Technology", KSG Faculty Research Working Paper Series, RWP08-009.

Keefe, Ryan, James P. Griffin and John D. Graham (2008). "The Benefits and Costs of New Fuels and Engines for Light-Duty Vehicles in the United States", Risk Analysis, 28, No. 5, 1141-1154.

Lazzari, Salvatore (2006). Tax Credits for Hybrid Vehicles, CRS Report for Congress, RS22558.

Small, Kenneth and Kurt Van Dender (2007). "Fuel Efficiency and Motor Vehicle Travel: The Declining Rebound Effect", Energy Journal, vol. 28, no.1, pp. 25-51.

Yacobucci, B.D., 2005. Tax incentives for alternative fuel and advanced technology vehicles, CRS Report for Congress, RS22351.

Heavy-Duty Vehicle Retrofits

Policy: Retrofits to heavy-duty vehicles can improve their fuel economy and thereby reduce emissions. Effective retrofits include changes to the tires to reduce roll resistance and changes to the body to reduce drag. States can pass regulations requiring the use of retrofits, or subsidize retrofits to voluntarily encourage their use.

Emissions Benefits and Costs: Retrofits can produce a 20%-60% improvement in fuel economy for individual vehicles, but aggregate effect depends on the percentage of existing vehicles incorporating retrofits. Costs to the public and to agencies are minimal since operators undertake retrofits. There are likely to be net social gains because the initial expense of retrofits can be recovered in fuel savings.

Implementation Concerns: Retrofits require up-front costs from operators, which they may resist, particularly in lean times, despite evidence that suggests that payback periods can be short.

Background

Heavy trucks are the preferred mode for short-to-medium distance freight transportation and for time-sensitive goods, and they are therefore a significant component of domestic freight movement. Heavy trucks have consumed nearly 33 billion gallons of gasoline, diesel, and other fuels annually in the U.S. in recent years, and virtually all of this fuel is derived from petroleum. This comprises approximately 18% of energy consumption in the transportation sector (Bureau of Transportation Statistics, 2009) and comes at the cost of associated greenhouse gas emissions, as well as more conventional pollutants.

Given the significant share of heavy-truck oil consumption, regulatory agencies (most notably the California Air Resources Board) are beginning to consider ways to improve the fuel efficiency of heavy trucks. One method to increase fuel efficiency is to target new truck sales with fuel economy standards. While this is likely to improve fuel efficiency over a two-decade time period, such a policy would not improve the average efficiency in the near term because of the slow rate of vehicle turnover.

A second policy option is to mandate retrofits to all heavy trucks to improve fuel economy. The diesel engines and powertrains in heavy trucks are not promising candidates for a retrofit strategy because of technical difficulties and the high costs of modifying existing engines and powertrains.[130] However, there are a variety of tire and vehicle body retrofits that can improve roll resistance and

---

[130] Retrofitting older diesel engines and adding new powertrain technology to older vehicles is not promising because there is a high cost to achieve small improvements in fuel economy. However, there are some minor engine adjustments that are designed to increase cylinder pressure or reduce internal friction that can cost-effectively improve the fuel economy of older heavy-duty trucks.

aerodynamics and present feasible and cost-effective opportunities for achieving greater fuel economy.

Policy and Implementing Organizations

State DOTs and environmental regulatory agencies can provide financial incentives for aerodynamic retrofits and/or advanced low-resistance tires, or implement regulations that mandate use of retrofits.[131] The EPA SmartWay program offers guidance, financial assistance, and other resources to increase freight efficiency.[132]

Effects

Target Group

Owners of heavy-duty truck fleets (and smaller independent companies and contractors) would be affected by retrofit incentives and/or mandates.

GHG Effects

Retrofit policies can encourage or require operators to either make specific modifications to existing heavy vehicles or choose retrofits. Historically, the largest concern about heavy-truck traffic has been the emissions of fine particulates and oxides of nitrogen, which have been tied to adverse health outcomes in urban areas and among sensitive populations (Dierkers et al., 2007). Thus, much of the research on retrofits focuses on conventional pollutants. More recently, studies have sought to understand how retrofits can reduce GHG emissions as well. There is a tendency in these studies to focus on the design of retrofits and their fuel and cost savings, rather than on the policy instruments to encourage their use—the assumption in most cases is that regulations will mandate their adoption.

Although different types of trucks can utilize and benefit differently from various types of retrofits, this section presents data for tractor trailers, since they represent the vast majority of heavy truck fuel consumption. While fewer options exist for "straight trucks," the efficiency effects are comparable. Effectiveness varies with assumptions about length of average trip (short, medium, long) and the number of trailers to which flaring technology (used to streamline the profile of trailers to reduce aerodynamic drag) is applied. The description below assumes a single trailer (to each cab) and the cost/effectiveness estimates below will reflect this assumption, as well as assumptions about the average trip length.

The Argonne National Lab study (Vyas, 2002) has estimated the following:

---

[131] The California Air Resources Board, for example, adopted new regulations in 2008 requiring certain types of heavy duty tractor trailers through aerodynamic and low rolling resistance tire retrofits, http://www.arb.ca.gov/cc/hdghg/hdghg.htm.

[132] http://www.epa.gov/smartwaylogistics/.

Retrofit Action	Fuel Economy Gain (%)	Cost ($) [133]
Cab top deflector	2.0	900
Gap closing	2.5	1,800
Trailer edge curvature	1.3	600
Pneumatic blowing	5.0	3,000
Low Rolling Resistance tires	3.0	1,300
Engine friction reduction	2.0	600
Peak cylinder pressure	4.0	1,200
Improved injectors	6.0	1,800
Vehicle mass (e.g., aluminum wheels)	5.0	2,400

Langer (2004) considered the cost-effectiveness of these strategies in a variety of scenarios with different fuel prices, discount rates for fuel savings, and payback periods. Langer (2004) found that currently available technologies provided net savings under all of these scenarios. Higher fuel prices will increase expected cost savings. Findings suggest that combinations of currently available technology retrofits lead to fuel economy improvements between 18% and 29% at a net savings of $2,500 to $15,000. Looking out to technologies available in 2015 or later, fuel economy improvements could be as high as 58%, and offer savings up to $24,000.[134] The results for more aggressive and costly retrofits like hybridization are mixed and the findings are dependent on the assumptions about fuel prices and payback periods (where lower fuel prices and shorter payback periods increase the net cost of the retrofit).

Schubert and Komer (2008) considered several retrofit packages ranging from a minimal package to an aggressive package. The minimal package consists of low rolling resistance tires and aerodynamic modifications to the cab. The aggressive package adds to that aluminum wheels (for weight reduction) and aerodynamic fairings to the trailer. Both packages use cost estimates for

---

[133] It is estimated that original values were reported in 2001 USD, which have been adjusted for inflation.

[134] The year in which these costs are calculated is not clear, so costs are documented as they are in the report, without converting to 2009 USD.

each modification that are comparable to the Argonne National Labs study above. Schubert and Komer (2008) considered scenarios of multiple fuel price, average trip length, and payback periods as well, and they find positive net savings in all but one case—and these savings are typically of several thousand discounted dollars.

Aggregate GHG effects depend on the penetration rates of new technology and the total area over which GHG savings are calculated. A state regulation that forces all vehicles in the state to adapt would lead to fuel savings everywhere the trucks operate, not just in the state that adopts this strategy. By 2020, after the California retrofit technology regulations are fully phased in, Schubert and Komer (2008) estimate that approximately 11–17 million $MTCO_2e$ will be saved annually. Cumulative GHG savings over twenty years are considerably higher, between 80-140 million $MTCO_2e$. Wider implementation of heavy-truck fuel economy or retrofit mandates could increase the effectiveness significantly, as heavy-trucks in more states would be forced to make retrofits.

Estimated Cost per Metric Ton of CO_2 Reduction

If one assumes that regulatory agencies or state/national governments bear no cost for implementing the regulations, then one may wish to consider the costs to fleet operators. Under most scenarios, the cost per metric ton of CO_2 savings is negative—i.e., results in a net savings—and under some scenarios, largely negative for operators, given the economic benefits of saving fuel.

Key Assumptions and Uncertainties

Access to capital remains an important factor in the penetration rates of these new technologies. Similarly, the cost of diesel fuel affects the net savings and necessary payback period.

Data and Tools

The Transportation Data Energy Book contains annual estimates of heavy-truck VMT and fuel consumption by state.[135] The EPA SmartWay website offers an extensive list of tools and data to help reduce emissions from heavy-duty trucks.[136]

Implementation Concerns

Agency Cost
Agency costs depend entirely on monitoring costs (unknown) and the decision to provide financial incentives for retrofits.

Agency Implementation Concerns
State DOTs and MPOs may have little authority to mandate changes to vehicles sold or registered outside of their jurisdictions.

---

[135] http://www-cta.ornl.gov/data/Index.shtml.

[136] http://www.epa.gov/smartway.

Social Concerns

Trucking companies are likely to oppose measures that increase operating costs in lean times—even if payback periods are relatively short. Without regulations, rapid penetration of these technologies in existing trucks seems unlikely, though they may become more common in new vehicles.

Other Costs/Barriers

None that are known to us at this time.

Interactions with Other Strategies

- Eco-driving education for long-haul truckers may shift social and industry preferences for fuel efficient trucks.

- Anti-idling regulations may offer some opportunities to combine fuel efficiency retrofits with those that enable anti-idling to reduce labor and materials costs.

Unique Co-benefits

None identified.

Negative Impacts

- The large upfront costs of the retrofits may be difficult for smaller independent operations.

Where in Use

California's heavy-truck fuel economy regulations are began phasing in starting in 2010, and ramp up until 2020 (CARB, 2009). Currently, no other state has comparable measures. From 2008–2010, the EPA's SmartWay program included approximately $50M to finance and help incentivize fuel-saving and emissions-reducing technologies.[137]

Recommendations for Further Research

It may be beneficial and relatively straightforward to model the implementation of statewide regulations to improve truck efficiency or mandate certain retrofits. However, California (and eventually other states) will need to clarify the penalties associated with non-compliance in order to fully understand the effectiveness of a retrofit policy.

References

California Air Resources Board (2009), AB32 Approved Scoping Plan: Appendices.

---

[137] http://www.epa.gov/smartway/financing/index.htm.

Dierkers, Greg, Mark Houdashelt, Erin Silsbe, Shayna Stott, Steve Winkelman and Mac Wubben (2007). CCAP Transportation Emissions Guidebook Part Two: Vehicle Technology and Fuels, Center for Clean Air Policy.

Langer, Terese (2004). Energy Savings Through Increased Fuel Economy for Heavy-Duty Trucks, American Council for an Energy Efficient Economy.

Schubert, Raymond and Matt Kromer (2008). Heavy-Duty Truck Retrofit Technology: Assessment and Regulatory Approach, Union of Concerned Scientists.

Vyas, A., C. Saricks, and F. Stodolsky (2002). The Potential Effect of Future Energy-Efficiency and Emissions-Improving Technologies on Fuel Consumption of Heavy Trucks, Center for Transportation Research, Argonne National Laboratory.

Eco-Driving Education and Training and Dynamic Eco-Driving

Policy: Small changes in driving behavior, collectively called "eco-driving," can improve fuel economy and reduce greenhouse gas emissions. Transportation agencies can encourage and enable eco-driving practices. Such changes include gentler braking and acceleration, slower driving, and avoiding idling. Three related approaches are education campaigns about eco-driving, eco-driving training programs that give drivers in-vehicle training, and dynamic eco-driving, which uses in-vehicle or road-based sensors to provide drivers with feedback about their behaviors and emissions.

Emissions Benefits and Costs: Because research on eco-driving impacts is still limited, it is difficult to draw definitive conclusions about emissions benefits and costs. Further, the effects of eco-driving depend upon the behaviors that are considered. Modest eco-driving can improve fuel economy by about 5%, although benefits can approach 30% in some circumstances. However, the effects of campaigns are largely unknown because few have been studied rigorously despite some evidence that it is one of the most cost effective ways to reduce GHG (one estimate suggests costs as low as $14/MTCO$_2$). Training programs can encourage drivers to adopt practices initially that produce a 5-15% improvement in fuel economy, but these gains tend to decline over time as drivers revert to earlier driving habits. The effects of dynamic eco-driving are as yet unknown as these programs have not been widely implemented. The cost effectiveness of training programs and dynamic eco-driving programs is also not known.

Implementation Concerns: There are no significant barriers to implementing eco-driving campaigns, training programs, and technology programs given that eco-driving is voluntary and campaigns can be low-cost. However, persuading people to eco-drive and therefore to achieve the outcomes of these programs may be more difficult.

Background

Small changes in driving behavior, collectively called "eco-driving," can improve fuel economy and reduce greenhouse gas emissions. Eco-driving on the road includes gentler braking and acceleration, slower driving, avoiding idling, driving in the highest gear, and using automated toll passes.

There are three related approaches to encouraging eco-driving that can be used individually or in concert:

1. Education campaigns about eco-driving, such as EcoDrivingUSA, provide informational material and literature on how to eco-drive and related benefits.

2. Eco-driver training programs give drivers in-vehicle training, sometimes on a course or at a driving school.

3. Dynamic eco-driving uses in-vehicle or road-based sensors to provide drivers with real-time or near real-time feedback on driving behaviors and emissions.

The relative effectiveness of these techniques is discussed in the "GHG Effects" subsection below.

Policy and Implementing Organizations

State and local transportation or other agencies and public-interest organizations may conduct eco-driving education and training campaigns independently or as part of traditional driver education programs. They may also provide incentives to install dynamic eco-driving sensors on vehicles. Private driver training programs have begun to include eco-driving in their course offerings, and industries that rely on heavy-duty vehicles in particular have sought to use eco-driving training to improve driving performance and reduce costs.

Effects

Target Group

Eco-driving campaigns and programs can potentially target all drivers.

GHG Effects

There is some quantitative research on training programs and dynamic eco-driving, mostly from Europe and involving heavy-duty vehicles, and there is general consensus about the directional effects when individuals adopt eco-driving. However, there is little quantitative research on eco-driving campaigns, in part because it is very difficult to measure such programs' effects.

According to the U.S. Department of Energy, eco-driving techniques such as driving sensibly, observing speed limits and removing excess weight, have the potential to improve a personal vehicle's fuel economy by between 5% and 33%.[138] Other sources, which may consider a different set of eco-driving techniques, report a 10% improvement (Barkenbus, in press).

Research on eco-driving training programs (where drivers learn and practice techniques on a driving course) and on eco-driving information technologies (in-vehicle or on-road) collectively suggest an improvement ranging from 5%-20%, with many studies reporting approximately 10%.

A review of the short- and long-term effects of eco-driver training programs in several countries suggests that, immediately after training, fuel economy from better driving styles can improve by 5-15%. In the midterm (approximately three years), fuel economy is approximately 5% better if no additional training is provided, and 10% if further training is provided (Workshop on EcoDriving, n.d.). This is generally consistent with results of the US Department of Transportation's eco-driving efforts of the 1970s (Greene, 1986). However, it is important to note that fuel prices are much higher in many parts of the world, notably Europe, than in the US. Thus, there may be less incentive

---

[138] http://www.fueleconomy.gov/feg/driveHabits.shtml.

in the U.S. to adopt eco-driving practices, and the findings from foreign studies may not be indicative of potential results in the U.S.

Research on the effects of dynamic eco-driving devices and technology is still emerging and conclusions are less certain. Driving simulator experiments and vehicle trials of an in-vehicle real-time eco-driving device suggested 16% and 11% reductions in fuel consumption, respectively, in comparison to driving without such a device (van der Voort, Dougherty, and van Maarseveena, 2001). Simulations of an on-road eco-driving system that encourages drivers to drive at more efficient speed suggested 25% reduction in CO2 emissions for the entire traffic stream in highly congested traffic, if 20% of drivers adopt the suggested speed (Barth and Boriboonsomsin, 2009). There are few real-world implementations of these devices. Pilot projects in Denver, CO and in the UK suggest 10% reduction in fuel consumption (Enviance, 2009; Greenroad, 2009). One pilot study in Southern California of an on-board eco-driving device that provided instantaneous fuel economy feedback showed that the city average fuel economy improved by 6% while highway economy improved by 1%. The study further found that participants are willing to adopt eco-driving practices, particularly at higher gas prices.[139]

To compute effects on greenhouse gas emissions, let us assume as the data suggests that these strategies improve a driver's driving habits such that a 10% improvement in fuel economy is achieved. Then, using data on average fuel economy and the number of vehicle miles traveled per vehicle (as a proxy for the number of vehicle miles traveled per driver), one can approximate the CO_2 emissions reduced as a result of eco-driving:

- For passenger cars: Using 2008 values, at 11,800 miles traveled per vehicle per year (National Transportation Statistics, 2009), an improvement from 22.6 MPG (the average fuel economy for passenger vehicles) to 24.86 MPG (a 10% improvement) results in approximately 47 gallons of gasoline saved and 930 lbs or 0.4 metric tons of CO_2 reduced annually per person.

- For light trucks: Using 2008 values, at 11,000 miles traveled per vehicle per year (National Transportation Statistics, 2009), an improvement from 18.1 MPG (the average fuel economy for passenger vehicles) to 19.91 MPG (a 10% improvement) results in approximately 55 gallons of fuel saved and 1,080 lbs or 0.5 metric tons of CO_2 reduced annually per person.

The effects of eco-driving education campaigns are much less certain because of difficulties in measuring changes in driving behavior and in attributing them to the effects of the campaign. Nevertheless, a long-running and aggressive national eco-driving campaign the Netherlands, with a total population of approximately 16.4 million people, 10 million licensed drivers, and 87 million VMT (Institute for Road Safety Research, n.d.), has saved approximately 0.3 MMT of CO_2 annually.

---

[139] http://trid.trb.org/view.aspx?id=1090479.

These reductions are increasing as more people are exposed to the program (Evaluation Dutch national ecodriving programme *Het Nieuwe Rijden* 2007, n.d.).

Estimated Cost per Metric Ton of CO_2 Reduction

In theory, eco-driving campaigns (e.g., formal public education and outreach on the nature and benefits of eco-driving) and programs may be among the more cost-effective ways to address GHG emissions (Barkenbus, n.d.), although costs will vary widely from program to program. In The Netherlands, the Dutch education program reported a program cost of $14 per metric ton of CO_2 (Wilbers, Wismans, and Jansen, 2004).[140] The costs associated with training-only or dynamic eco-driving systems are not currently known.

Key Assumptions and Uncertainties

There are uncertainties in the long-term effects of eco-driving training and also the extent to which campaigns can reach the public and alter behavior. There is also uncertainty and variation in the cost of campaigns.

Data and Tools

None identified.

Implementation Concerns

Agency Cost

Costs to transportation agencies from undertaking eco-driving programs depend on the extent of campaign activities they undertake. Sources of cost include signage and postings, employee salaries, advertising costs, and training systems and technologies.

Agency Implementation Concerns

There are no specific agency implementation concerns associated with eco-driving education and training programs. However, agencies' eco-driving programs may be more effective if they are implemented in cooperation with automobile associations, whose members number in the millions.

Social Concerns

The social acceptability of such programs is likely to be high given that they are voluntary. Expenditures by individuals, government, and industry for eco-driving education and training will likely be negligible at a national level. As eco-driving technology is as yet largely undeveloped, the costs to individuals, government and industry for such technology is unknown but conceivably would be implemented as part of larger transportation infrastructure projects and ongoing vehicle technology advancements.

---

[140] This cost was originally reported as 10EUR per metric ton of CO_2 in 2007; it has been converted to 2009 USD.

Other Costs/Barriers
None identified.

Interaction with Other Strategies
- Eco-driving programs can be used alone and while they are likely to benefit other strategies (e.g., anti-idling regulations) by increasing their social acceptability, it is not clear how other strategies would specifically improve eco-driving programs.

Unique Co-benefits
- There may be safety benefits associated with less aggressive driving.

Unique Negative Effects
None identified.

Where in Use
Eco-Driving campaigns are more common in the EU than in the US. EU programs exist in The Netherlands, Sweden, Portugal, and the UK. In the U.S., the Eco-Driving USA campaign is national and endorsed by many state governors, but it is not clear that any state or region has undertaken its own campaign. Some higher-end vehicles also already provide instantaneous fuel economy feedback.

Recommendations for Further Research
In the longer term, eco-driving education and training programs should be undertaken and/or studied in the US to provide more information on the effects of such programs. There may also be value in conducting long-term studies of the impact of in-vehicle eco-driving systems.

References
Barkenbus, J.N. (in press). Eco-driving: An overlooked climate change initiative. *Energy Policy*. Available online 28 October 2009 at http://www.sciencedirect.com/science/article/B6V2W-4XJN4X4-5/2/6ffebae2ed6ed89c7f0c083fa7b37e6d.

Barth, M., and Boriboonsomsin, K. (2009). Energy and emissions impacts of a freeway-based dynamic eco-driving system, *Transportation Research Part D*, pp. 400—410.

EcoDrivingUSA (n.d.). The Ecodriver's manual: A guide to increasing your mileage and reducing your carbon footprint [brochure]. Available online at http://www.ecodrivingusa.com/files/EcoDriving_Manual.pdf.

Enviance. (2009, January 27). *Denver's driving change program reduces vehicular CO_2 emissions* [press release]. Available online at http://www.enviance.com/about-enviance/PressReleaseView.aspx?id=53.

Evaluation Dutch national ecodriving programme Het Nieuwe Rijden 2007 (n.d.). [Brochure]. Available at http://fiabrussels.com/download/projects/ecodriven/factsheet_evaluation_hnr_2007_en2.pdf.

Greene, D.L. (1986). *Driver energy conservation awareness training: review and recommendations for a national program.* Report from the Oak Ridge National Laboratory (No. ORNL/TM-9897).

Greenroad. (2009). *Datashred's fleet goes green.* Available online at: http://www.greenroad.com/success_story_phs.html.

Institute for Road Safety Research (n.d.) - http://www.swov.nl/cognos/cgi-bin/ppdscgi.exe?BZ=1AAABegNwdgUABEwU6VFChhEnZciUkcMmjBsyc59ZCSMnTdg~UtKoEVPmbhtz29GTho4JIVF6kIEBI2Z3fxskSJDsq9klszvH5taeWUwiZiURs9hESDApU8YM3eaQKQEjh5Kw9fymSZEgU6pIKTL1tZPB2TUDZl_ZmH1iu3Jqg5hEAxJSryTzJ8ekp4zbw41724KRxs2dN25L6178hiMGTkm0doZCLRenUMt16xKHmF5mXJP0mT~gHz==). OR http://www.swov.nl/uk/research/kennisbank/inhoud/90_gegevensbronnen/inhoud/data_2.htm.

Van der voort, M., Dougherty, M.S., and Van Maarseveee, M. (2001). A prototype fuel-efficiency support tool, *Transportation Research Part C*, 279—296.

Workshop on Eco-Driving: Findings and Messages for Policymakers (n.d.). [Report from workshop, Paris, November 22-23, 2007]. Available online at: http://www.internationaltransportforum.org/Proceedings/ecodriving/EcoConclus.pdf.

Truck Stop Electrification and Auxiliary Power Units

Policy: Truck-stop electrification (TSE) and auxiliary power unit (APU) technologies provide long-haul truckers with heating, cooling, and other amenities at truck stops without requiring vehicle idling, thereby reducing GHG emissions. Agencies can encourage the adoption of TSE and APUs through funding and partnerships with private companies.

Emissions Benefits and Costs: Using TSEs or APUs instead of idling reduces GHG emissions by 60% or more. The aggregate effects depend on the number of hours of idling that are actually offset. Without considering revenue generated from providing power services, the cost is $20-$60/MTCO$_2$ for TSE systems, depending on usage rates and system lifespan. The cost of APUs, on the other hand, can be fully recovered by operators in 2-3 years from lower fuel and maintenance costs.

Implementation Concerns: TSE offers business opportunities to truck stop operators and TSE and APUs both reduce costs for fleet operators. Nevertheless, acquiring financing for APUs and other technologies may pose a barrier for fleet operators. The cost to public agencies depends on the level of support they choose to offer.

Background

Federal safety regulations require that truckers must rest ten hours for every eleven hours of consecutive driving (Federal Motor Carrier Safety Administration, n.d.). In complying with these regulations, long-haul truck drivers idle their engines 5-8 hours a day to power air conditioning, heat, and other on-board appliances, and to keep engines and fuels warm in cold weather. Trucks typically consume 0.8 gallons of diesel fuel per hour of idling, using between 900 and 1,400 gallons of fuel each year per truck. This extensive idling results in significant GHG emissions.[141]

Truck stop electrification (TSE) technologies reduce extended idling at truck stops by providing electricity-powered heating, cooling, and other amenities. On-board TSE solutions require some vehicle modification and use batteries on the truck to power appliances, and they offer outlets at truck stops to recharge these batteries. Off-board TSE solutions (which require no vehicle modifications) provide complete heating and air conditioning infrastructure via an overhead unit and a hose connection. In addition to basic heating and cooling, off-board systems can offer Internet access, movies, and satellite programming (California Energy Commission, 2006). These options generate revenue for truck stop operators and simultaneously are less costly to truck operators than idling because of lower electricity costs (relative to diesel fuel costs for the same energy) and lower maintenance costs that would be incurred because of the negative effects of idling on the engine.

---

[141] http://www.epa.gov/smartway/documents/partnership/trucks/partnership/techsheets-truck/EPA420F09-038.pdf.

U.S. Department of Transportation
Federal Highway Administration

Other anti-idling technologies include auxiliary power units (APUs), which typically provide heat and electricity through small, externally mounted, diesel-powered internal combustion engines, and cooling through electric air conditioners or the vehicle's air conditioning system. APUs are proven, commercially available technologies and are efficient because the engine is appropriately sized to meet heating and electricity needs (Argonne National Laboratory, 2000).

Policy and Implementing Organizations

State DOTs, MPOs, and other agencies (e.g., state environmental protection or energy agencies) can provide funding and strategic planning support to truck stop operators and truck operators to implement on-board and off-board TSE and APUs. Such projects are frequently undertaken as public-private partnerships and agencies or operators may seek funding and support from other (e.g., federal) sources. The EPA SmartWay program offers guidance, financial assistance, and other resources to freight operators for using such technologies.

Effects

Target Group

TSE and APU projects are targeted at truck stop operators, long-haul truckers, and fleet operators. By providing funding to truck stop operators, these efforts seek first to enable and encourage truck stop operators to install TSE facilities. Second, assuming TSE facilities can offer amenities at lower costs to truckers than burning diesel fuel, these efforts aim to enable and encourage truckers to use TSE facilities. TSE and APU projects can also enable fleet operators to install on-board TSE or APU equipment.

GHG Effects

Public agencies have been successful in enabling TSE projects through financial and strategic planning support. However, it is not possible to generalize and quantify the effects of funding opportunities on operators' decisions to undertake TSE projects, since these efforts tend to be public-private partnerships and the decision to implement TSE depends on a wide range of factors. These factors include the total cost of implementation, the level of funding available from public sources, the demand from truckers and fleet operators, and the anticipated revenue to operators.

Nevertheless, theoretical and practical studies do provide estimates of the reductions in emissions from using TSE or APUs instead of idling. There is some variation in these estimates at all levels (hourly, yearly, per-space, and per-site) due to different assumptions about and variations in power requirements, fuel efficiency, facility usage. The literature also includes estimates of system implementation costs and costs per metric ton of GHG reduction. Again, there are some variations depending on the type of technology used and assumptions about technology lifespan. Many TSE projects in particular are less than five years old, so actual long-term costs and benefits are unknown.

As noted, long-haul truck drivers idle their engines five to eight hours a day to power air conditioning, heat, and other on-board appliances, and to keep engines and fuels warm in cold weather. Trucks typically consume 0.8 gallons of diesel fuel per hour of idling. Given that one gallon of diesel fuel emits 22.4 lbs of CO_2, then an idling truck would emit approximately 18 lbs of CO_2 per hour or 90-145 lbs of CO_2 per day.[142]

In contrast, if an hour of off-board TSE use draws 7.5kW (Center for Clean Air Policy, 2007), and given an average emission of 1.33 lb of CO_2 per kWh from the electricity grid (EPA, 2008), then a truck that uses a TSE spot would produce approximately 10 lbs of CO_2 per hour, or approximately 60-80 lbs of CO_2 per day.

Other reports use different estimates of both the fuel consumed for idling and the equivalent electricity requirements:

- A report from the Argonne National Laboratories (2000) found that one hour of idling produces 21.8 lbs of CO_2 (131-174 lbs per day) but that one hour of off-board TSE use requires only 4.3kW and emits only 6.3 lbs of CO_2 per hour (38-51 lbs of CO_2 per day). It also found that APUs emit only 4.1 lbs of CO_2 per hour (25-33 lbs of CO_2/day). While on-board units add weight, this weight is small enough (less than 0.4% of the vehicle's overall weight) as to have no effect on the fuel economy.

- A more recent study from the Argonne National Laboratories (Gaines and Hartman, 2009) estimated that idling emits 20 and 15 lbs of CO_2 per hour for 2001 and 2007 truck model years, respectively, while TSE emits only about 3 lbs per hour[143] and APUs emit 6 lbs of CO_2 per hour. These findings differed slightly from the earlier study, which suggested that APUs had lower emissions than TSE, though the actual figures are comparable.[144] This results in differences between 90-160 lbs of CO_2 per day from idling versus 19-26 lbs of CO_2 per day from TSE and 36-42 lbs of CO_2 per day from APU use.

In sum, this amounts to approximate a 60% or more decrease in GHG emissions from idling to TSE or APU use.

The absolute effect on GHG emissions from an individual APU or TSE spot depends on the number of hours of idling that have been offset. Because APUs are installed in individual vehicles, variations in use are largely a function of travel patterns and climate. TSEs on the other hand, are designed to

---

[142] http://www.epa.gov/smartway/documents/partnership/trucks/partnership/techsheets-truck/EPA420F09-038.pdf.

[143] This study does not distinguish between on-board and off-board TSE, and instead only considers the electrical load. Additionally, it reports annual emissions per vehicle and annual idling hours per vehicle, which this sourcebook has used to compute emissions per hour.

[144] The differences in these figures can, in part, be traced back to different estimates of energy use for the different systems. The 2000 study reports 23,000 and 45,000 BTUs for APUs and TSE, respectively, while the 2009 study reports approximately 35,000 and 10,000 BTUs. The reasons for these differences are not clear.

serve many vehicles, and their use depends additionally on their location, number, and distribution of nearby TSE facilities, and other factors. Reports from TSE proposals and projects illustrate some of the benefits observed or anticipated:

- The final report of several TSE sites in Pennsylvania (Shulman, 2008) estimated that the facility in Carlisle, PA with 72 TSE spots reduced CO_2 by 3,450 metric tons in 18 months of operation. Note that this project originally estimated 60% utilization of the TSE system but, in practice, has achieved only about 35% utilization.[145]

- A TSE proposal from the Georgia Environmental Protection Division (Cook, undated) estimated that the proposed 85 off-board TSE spots would reduce 2000 tons[146] of CO_2 emissions per year, assuming utilization of 10 hours per space per day.

- A request for proposals for TSE projects in Oregon (as cited in Downing, 2005) from the Climate Trust estimated that the 600 electrified stops would reduce CO_2 emissions by 33,000 tons annually.[147]

- A demonstration project involving 18 on-board TSE spaces along New York's I-87 determined that over 13 months, these spaces saved 393 metric tons of CO_2. While this outcome is low relative to other TSE projects, this may be explained partly by the facts that only 10% of trucks have the necessary on-board equipment installed, direct marketing was minimal, and that this facility was, at the time, the only such site on the East Coast (Antares Group Inc., 2005).

A 2000 report found that if market penetration of TSE or APUs was 100%, 7 to 8 $MMTCO_2$ could be saved annually (Argonne National Laboratories, 2000).

Estimated Cost per Metric Ton of CO_2 Reduction

- Implementation costs for a single off-board TSE spot can vary substantially. Stated estimates of off-board systems range from $7,000 (EPA, n.d.) to $25,000 (Antares, 2005).[148] One can use data from project proposals and reports to approximate costs, though assumptions must be made about utilization rates over time and the lifespan of TSE spots.

---

[145] The report did not clarify the reasons that actual use was significantly below projected use.

[146] It is unclear from the report whether data is in short tons or metric tons.

[147] It is unclear from the report whether data is in short tons or metric tons.

[148] None of the sources cited in this section state the year for which costs were calculated; the costs are thus cited as they were noted in the reports. In addition, many sources did not specify whether reductions were measured in short or metric tons; where there is ambiguity, the data are cited simply in "tons" as was done in the reports.

The proposal from the Georgia Environmental Protection Division (undated) stated a cost of $37 per ton of CO_2.

- A final report of Pennsylvania's TSE project (Shulman, 2008) estimated that 72 spots reduced 3450 metric tons of CO_2 in 18 months, or 32 metric tons per space per year. The total cost for a single spot varied from between $15,000 and $19,000 per spot. Assuming a 10-year lifespan per space (as suggested by the proposal from the Georgia Environmental Protection Division [undated]) and no change in utilization rates, this would result in a cost between $47 and $60 per metric ton of CO_2.

- A request for proposals for TSE in Oregon (as cited in Downing, 2005) anticipated spending $7 million to implement 600 electrified stops ($11,000 each) and expected a total annual reduction of 33,000 tons of CO_2. Assuming a 10-year lifespan again, this results in an estimated cost of between $21-$23 per ton of CO_2 (depending on whether reductions are estimated in short or metric tons).

On-board TSE spots cost less to implement at the truck stop but require equipment installation in trucks. The study of New York's on-board TSE (Antares Group, Inc., 2005) estimated that the costs to install one on-board TSE spot ranges from $3,000 to $6,000 and that 363 metric tons of CO_2 were reduced in 13 months. Continuing with our assumption of a 10-year lifespan and assuming the same utilization rates, the cost per metric ton of CO_2 ranges from $15 to $30 per metric ton. Note that for truck stop operators, this cost is likely to be offset by revenues from vehicle operators who use the TSE. This cost figure does not include the cost for on-board equipment, which ranges from $180 to over $3000, but which can be recovered by operators in a few years from fuel savings (Argonne National Laboratories, 2000).

Operators can recover the $6,000 to $7,000 cost of APUs in less than two years from fuel and engine maintenance savings (Argonne National Laboratories, 2000).

Key Assumptions and Uncertainties

There is significant uncertainty in the actual rates of use of TSE spots and the change in use over time. Variations in fleet efficiency, climate, CO_2 emissions from electricity generation, and lifetime costs of TSE further contribute to uncertainty in both the costs and level of reductions.

Data and Tools

- The EPA publication "Guidance for Quantifying and Using Long Duration Truck Idling Emission Reductions in State Implementation Plans and Transportation Conformity" provides information on calculating the benefits of technologies including truck stop electrification. This can be accessed at http://www.epa.gov/smartway/documents/420b04001.pdf.

- The EPA's SmartWay website provides general information on anti-idling technologies (http://www.epa.gov/smartway/).

- EPA's MOVES model can be used to model the effects of reductions in long-haul truck extended idling.

Implementation Concerns

Agency Cost

TSE and APU project costs stem from planning, technology implementation, outreach, and evaluation activities. The cost to agencies depends entirely on the nature and size of support they offer to operators. For example, the state of Pennsylvania contributed $900,000 for three TSE sites, while the EPA contributed $100,000 and the private partner, IdleAire, contributed $2.5M (Shulman, 2008).

Agency Implementation Concerns

TSE projects can involve partnerships between state and local transportation agencies, other state and local organizations, federal agencies, and private companies, and such complex partnerships present collaborative challenges. There are no apparent significant inter-agency or institutional concerns associated with APU use at this time.

Social Concerns

TSE facilities are socially viable because of the business opportunities they provide to truck stop operators and TSE technology providers, and, as with APUs, because of the cost-saving opportunities they provide to truck operators.

Other Costs/Barriers

The installation of both on-board and off-board TSE facilities requires substantial initial investment from truck stop operators and other firms. On-board TSE systems also require investments from truck operators. However, public agencies can provide support and incentives to reduce these costs, and the success of such initiatives in generating TSE projects suggests that this initial hurdle can be overcome.

Despite the benefits that TSE offers to both truck stop operators and fleet operators, the development of TSE spots has been uneven. In 2009, when this work began, there were 138 TSE locations around the U.S. As of August 2010, only 12 locations remained because the company that owned nearly all TSE locations had filed for bankruptcy. The specific causes of this are not known, though the economic downturn in recent years may have played a role (e.g., in limiting investments in new infrastructure or technology).[149]

---

[149] This information was obtained from personal communication with the U.S. Department of Energy's (DOE) Clean Cities Technical Response Service in August 2010. At the time, this service indicated that TSE sites starting to reopen.

Interaction with Other Strategies

- TSE facilities and APUs can be implemented successfully on their own, but the use of these technologies could be increased through simultaneous eco-driving and anti-idling education programs targeted at truckers. Such programs would inform truckers of the drawbacks of idling, the availability of alternatives in the form of APUs and TSE facilities, and the financial and environmental benefits they offer.
- Virtually none of the existing anti-idling regulations applies to long-haul truckers, in recognition of the need to operate heating, cooling, and other needs or amenities by idling. However, such regulations may be a feasible way to increase APU use or TSE use once there are enough TSE facilities to essentially always offer truckers alternatives to idling.

Unique Co-benefits

- TSE and APUs may reduce costs for fleet operators while generating revenue for truck stop operators.

Unique Negative Effects

None identified.

Where in Use

A searchable map of TSE sites can be found at http://www.afdc.energy.gov/afdc/locator/tse/. There has been a significant decline in the number of TSE locations, from approximately 140 at the end of 2009 to only 12 in August 2010.

Recommendations for Further Research

The data on TSE projects is distributed in individual project proposals and reports. While this section cites data from a few of these, a broader survey of projects would provide more comprehensive data and insights into the costs and benefits of TSE.

References

Antares Group, Inc. (2005). *Fleet Demonstration of Shorepower Truck Electrified Parking On The I-87 Northway* [Final report for NYSERDA]. Retrieved from http://www.nyserda.org/publications/I87TSEDemonstrationReportDec05.pdf.

California Energy Commission (2006, June). *Truck Stop Electrification* (Publication No. CEC-600-2006-001-FS). Retrieved from http://www.energy.ca.gov/2006publications/CEC-600-2006-001/CEC-600-2006-001-FS.PDF.

Center for Clean Air Policy (2007). *CCAP Transportation Emissions Guidebook* [online], Washington, DC. Accessed on January 6, 2010 from http://www.ccap.org/safe/guidebook/guide_complete.html.

Cook, W. (n.d.). *Georgia Truck Stop Electrification (TSE) and Green Corridors* [Proposal from the Georgia Environmental Protection Division for American Recovery and Reinvestment Act Funding] Retrieved from http://www.gaepd.org/Files_PDF/arra/ARRA_truck_work_plan.pdf.

Downing, K. (2005, Winter). Saving Energy, The Environment, And A Good Night's Rest—Oregon's Approach To Truck Idling. *The Journal of the Environmental Council of the States*, 17-19. Retrieved from http://www.ecos.org/files/1411_file_Winter_2005_ECOStates.pdf.

Environmental Protection Agency (2004, January). *Guidance for Quantifying and Using Long Duration Truck Idling Emission Reductions in State Implementation Plans and Transportation Conformity* (Publication No. EPA420-B-04-001). Retrieved from http://www.epa.gov/smartway/documents/420b04001.pdf.

Environmental Protection Agency (2008, December). *eGRID2007 Version 1.1 Year 2005 Summary Tables*. Accessed on January 6, 2010 from http://www.epa.gov/RDEE/documents/egridzips/eGRID2007V1_1_year05_SummaryTables.pdf.

Environmental Protection Agency (n.d.). *Technologies, Strategies and Policies: Idling Reduction.* Accessed on January 6, 2010, from http://www.epa.gov/smartway/transport/what-smartway/idling-reduction-tech.htm.

Federal Motor Carrier Safety Administration (n.d.). *Interstate Truck Driver's Guide to Hours of Service*. Accessed on January 6, 2010 from http://www.fmcsa.dot.gov/rules-regulations/truck/driver/hos/fmcsa-guide-to-hos.PDF.

Gaines, L. and Hartman, C. (2009). Energy Use and Emissions Comparison of Idling Reduction Options for Heavy-Duty Diesel Trucks. In *Proceedings of the Annual Meeting of the Transportation Research Board,* Washington, D.C., Paper No. 09-3395. Retrieved from http://www.transportation.anl.gov/pdfs/TA/397.pdf.

Lim, H. (2002, October). *Study of exhaust emissions from idling heavy-duty diesel trucks and commercially available idle-reducing devices.* Environmental Protection Agency. (Publication No. EPA420-R-02-025). Retrieved from http://www.epa.gov/otaq/retrofit/documents/r02025.pdf.

Perrot, T., Dario, J., Kim, J., and Hagan, C. (2004, September) *Installation and Economics of a Shorepower Facility for Long-Haul Trucks* [Report for NYSERDA]. Retrieved from http://www.nyserda.org/publications/Shorepower.pdf.

Shulman, A. (2008). *Final Report Award # XA – 83207301-0* [Final report from the Pennsylvania Department of Environmental Protection to the US Environmental Protection Agency]. Retrieved from http://www.epa.gov/smartway/documents/adeq-pennsylvania-final-report.pdf.

Stodolsky, F., Gaines, L., and Vyas, A. (2000). Analysis Of Technology Options To Reduce The Fuel Consumption Of Idling Trucks (Technical Report ANL/ESD-43). Retrieved from Argonne National Laboratory Website: http://www.transportation.anl.gov/pdfs/TA/15.pdf.

U.S. Department of Energy, (2009). Energy Efficiency and Renewable Energy, Truck Stop Electrification for Heavy-Duty Trucks.
http://www.afdc.energy.gov/afdc/vehicles/idle_reduction_electrification.html.

Anti-Idling Regulations and Campaigns

Policy: Idling in traffic may be necessary for safety and system efficiency, but idling to warm the engine and idling while waiting for non-traffic reasons is generally unnecessary. Anti-idling regulations and campaigns seek to require or encourage drivers to reduce vehicle idling, thereby reducing greenhouse gas emissions.

Emissions Benefits and Costs: While the effects of idling are understood, it is not known how regulations and campaigns affect drivers' idling behaviors. Additionally, costs of anti-idling regulations are largely unknown.

Implementation Concerns: The public generally supports voluntary anti-idling campaigns, and opposition to regulations can presumably be overcome given that anti-idling regulations are widespread.

Background

Drivers of passenger vehicles and commercial trucks (excluding long-haul freight) typically idle their vehicles in three situations: while waiting in traffic (e.g., at traffic lights), while waiting for non-traffic reasons (e.g., waiting for other passengers or while making deliveries), and to warm the engine (Carrico, 2009). While idling in traffic may be necessary for safety and system efficiency, idling to warm the engine (excluding rest-periods for long-haul trucking) and idling while waiting for non-traffic reasons is generally unnecessary. It also results in wasted fuel, increases conventional air pollutants, and emits greenhouse gases.

In an effort to curb conventional air pollutant emissions in particular, many state and local governments have implemented anti-idling laws that limit idling times to various extents depending on the class of vehicle, zoning, time of day, and environmental conditions. Transportation agencies and public interest organizations have also undertaken anti-idling campaigns, separately or in conjunction with regulations, to educate drivers about the effects of idling. Despite the traditional focus on conventional pollutants, these regulations and campaigns simultaneously reduce greenhouse gas emissions.

Policy and Implementing Organizations

State and local governments are responsible for passing regulations that limit idling. The role of transportation agencies in implementing these policies is to provide appropriate signage. Law enforcement or parking/traffic enforcement agencies would enforce the regulations. Transportation agencies and public interest organizations may also undertake anti-idling education campaigns.

Effects

Target Group

Anti-idling regulations and campaigns can potentially affect all motorists and all vehicles. Some may be targeted at specific groups or areas, such as anti-idling regulations for school buses.

GHG Effects

Several studies have sought to quantify the extent of unnecessary idling of private and commercial vehicles. Light-duty vehicle idling emits approximately 1.4g of CO_2 per second (Frey et al., 2003), or 84 grams of CO_2 per minute. A 2007 survey of drivers by Carrico et al. (2009) suggested that the average passenger-car driver unnecessarily idles their vehicle for approximately 6 minutes every day, emitting 1.1 lbs of CO_2. This results in approximately 16 MMT of CO_2 annually in the U.S. with unnecessary engine warming and waiting.

A study by the Argonne National Laboratory (Gaines, Vyas, and Anderson, 2006) estimated that commercial vehicles emit between 40 and 170 grams of CO_2 per minute and that drivers of these vehicles idle from half an hour to two hours a day, depending on the type of vehicle and commercial activity. This translates to unnecessary emissions of between of over 20 lbs of CO_2 per day for certain vehicles. In sum, unnecessary commercial truck idling can consume 2.5 billion gallons of gas and emit approximately 23 MMT of CO_2 annually.

Although anti-idling regulations and campaigns have been implemented in various ways in many places to counteract such idling, there is little quantifiable data on the extent to which they actually change driver behavior, either on a per-person or on an aggregate level. The effectiveness of legislation depends in part on how the legislation itself is structured (schedule of penalties, exemptions to rules, etc.), how it is made known to the public (e.g., extent of signage and education campaigns), and how it is enforced (complaints from the public, enforcement blitzes, ongoing proactive enforcement). Not only is there variability in the regulations and their implementation (and hence their effects), there is also little evidence and data available about the effects of particular regulation. The reasons for this include:

- the difficulty of measuring before-and-after idling behaviors among the public as a whole;

- the difficulty of using trends in the number of idling citations issued as an indirect measures of public behavior because:

    - enforcement activities can be weak[150] and inconsistent over time;

    - it is difficult to attribute any changes to idling behaviors to regulations specifically, as behaviors are also influenced by gas prices, education campaigns, weather, or other factors; and,

- few governments report data on idling citations.

Many of these informational and data collection challenges are common to anti-idling campaigns, but some campaigns have made informal efforts to gather direct and indirect data for short periods of time. Some campaigns reported that between 50% and up to 95% of idling drivers approached

---

[150] New York City, for example, has a relatively stringent anti-idling regulation on the books (three minutes), but only 325 citations for violations were issued in 2003 and 526 in 2007 (New York City law cracks down on idling cars, 2009).

by campaigners complied with requests to limit idling and/or expressed a commitment to reduce idling (Kings County Economic Development Agency, n.d.; Freedman, 2009), but it is unclear if and how that translates into regular changes in behavior. Most campaigns that observed idling behavior reported inconclusive and mixed results (e.g., because of confounding effects of different weather conditions during data collection times) (Kings County Economic Development Agency, n.d.; Transport Canada, 2004).

Private companies and organizations may also undertake anti-idling campaigns and target a particular group of drivers (bus drivers, police fleet, etc.). A survey of several anti-idling training efforts conducted by various companies for their delivery truckers decreased emissions by 60 to 160 lbs of CO_2 per vehicle per week in the two weeks following training (Engine Idling - Costs You Money and Gets You Nowhere, n.d.), but it is unclear how these campaigns have affected long-term behavior.

Estimated Cost per Metric Ton of CO_2 Reduction

Costs to the public largely depend on the extent of the education and campaign activities, and enforcement activities in the case of regulations. The nature of these efforts and their costs vary. Given that both the effects and costs are unknown and highly variable, it is not feasible to calculate the cost per metric ton.

Key Assumptions and Uncertainties

There are significant uncertainties in the effect of campaigns and regulations on behavior, particularly given variation in policy structure and enforcement efforts (for regulations), and given that the effects are confounded with other influencing factors such as the cost of gas, anti-idling campaigns. There are also uncertainties and high variability associated with cost.

Data and Tools

EPA's MOVES model can be used to estimate changes in emissions.

Implementation Concerns

Agency Cost

Costs to transportation agencies for anti-idling regulations vary and depend in part on the approach they take for education and enforcement. Costs include signage and postings associated with the regulation, training of enforcement personnel, hiring new personnel, and creating a hotline for receiving public complaints. In some cases, policymakers may not anticipate any new costs associated with enforcement (Pennsylvania Department of Environmental Protection Environmental Quality Board, 2008).

Costs for campaigns largely depend on the extent of campaign activities, but are not likely to be expensive overall. The literature from several locations showed costs ranging from $4,000 for signage (Kings County Economic Development Agency, n.d.) to $130,000 for a more extensive media campaign (Transport Canada, 2004), but this is by no means an upper bound. This $130,000 cost included staff resources of $50,000 for one year, production costs of $30,000 and $50,000 for evaluation. Various organizations may undertake these campaigns and bear the costs.

Agency Implementation Concerns

There are no specific agency concerns associated with anti-idling regulations or campaigns, though their effectiveness depends on how and how well they are enforced. For example, a recent report found that New York City police rarely ticket for idling, despite regulations (New York City law cracks down on idling cars, 2009). In Canada, most municipalities with by-laws have taken a limited approach to enforcement. Typically, communities do some public outreach and education on the issue of vehicle idling prior to passing the by-law as well as afterwards. Enforcement has mainly involved reacting to complaints from the public by speaking to offenders, providing information on the by-law and the reasons for it, and asking for voluntary compliance. Few communities issue tickets and summonses, and those that do usually limit this activity to short sporadic campaigns rather than undertaking ongoing enforcement (Clean Air Partnership, 2005).

Social Concerns

Surveys conducted by campaigns suggest that the public supports voluntary anti-idling efforts for air quality purposes in particular, and acceptability for these campaigns is likely to be high. The literature does not discuss public support for or opposition to anti-idling regulations, but, given that just over half of all states have anti-idling regulations, one can infer that they are socially acceptable. Nevertheless, support likely varies in part on the environmental and health concerns of the public, who the regulation targets, and how.

Other Costs/Barriers

None identified.

Interactions with Other Strategies

- Anti-idling regulations and campaigns can be implemented on their own but may be more successful with eco-driving campaigns, which seek to increase awareness about the benefits of various driving behaviors.

Unique Co-benefits

None identified.

Unique Negative Effects

None identified.

Where in Use

As of 2006, approximately 30 states had some type of anti-idling regulation, either at the state level or in particular cities or counties. EPA provides a list at: http://www.epa.gov/smartway/documents/420b06004.pdf. Many regions have undertaken anti-idling campaigns at some point or another, but there is no comprehensive database or list of such efforts.

Recommendations for Further Research

Anti-idling campaigns and regulations are rarely evaluated, so further research would be helpful in designing and executing evaluations. In the near term, additional information about anti-idling regulations may be gleaned from individual states' planning documents, particularly related to costs of education campaigns and enforcement and related to the extent to which states have monitored the effects of such rule making (at minimum, in terms of citations, and more broadly, in terms of other behavioral effects).

References

Carrico, A., Padgett, P., Vandenbergh, M.P., Gilligan, J., Wallston, K.A. (2009, August). Costly myths: An analysis of idling beliefs and behavior in personal motor vehicles, *Energy Policy*, 37, Issue 8, 2881-2888.

Center for Clean Air Policy (2007). *CCAP Transportation Emissions Guidebook* [online], Washington, DC. Accessed on January 6, 2010 from http://www.ccap.org/safe/guidebook/guide_complete.html.

Freedman, R. (2009) Idle-Free Ambassador Program Evaluation [Project Report].

Frey, H.C., Unal, A., Rouphail, N.M., Colyar, J.D. (2003). On-road measurement of vehicle tailpipe emissions using a portable instrument. *Journal of Air and Waste Management Association*, 53, 992–1002.

Gaines, L., Vyas, A., and Anderson, J. L. (2006). Estimation of fuel use by idling commercial trucks. **Transportation Research Record: Journal of the Transportation Research Board**, 1983(-1):91-98.

Kings Community Economic Development Agency (n.d.). Kings County Anti-Idling Campaign Final Report.

Clean Air Partnership (2005). Cracking Down on Idling: A Primer for Canadian Municipalities on Developing and Enforcing Idle-free By-laws. Prepared for the National Resources Canada and the Greater Toronto Area Clean Air Council. http://www.crd.bc.ca/rte/idling.htm.

New York City law cracks down on idling cars. (2009, Feb 20). *MSNBC*. Retrieved from *http://www.msnbc.msn.com/id/29258343/wid/6448213/*.

Pennsylvania Department Of Environmental Protection Environmental Quality Board (2008). Regulatory Analysis of Diesel Vehicle Idling, 25 Pa. Code§ 121.1 and Chapter 126, Subchapter F.

Penney, J. (2005). *Situational Analysis: The Status of Anti-idling By-laws in Canada [Report prepared for Natural Resources Canada and The Greater Toronto Area Clean Air Council]*

Transport Canada (2004). *Towards an Idle-Free Zone* [online] on the Urban Transportation Showcase Program, Case Studies in Sustainable Transportation. http://www.tc.gc.ca/eng/programs/environment-utsp-towardsanidlefreezone-1076.htm.